SpringerBriefs in Materials

More information about this series at http://www.springer.com/series/10111

Naznin Sultana · Mohd Izzat Hassan
Mim Mim Lim

Composite Synthetic Scaffolds for Tissue Engineering and Regenerative Medicine

 Springer

Naznin Sultana
Mohd Izzat Hassan
Mim Mim Lim
Faculty of Biosciences
  and Medical Engineering
Department of Clinical Sciences
Universiti Teknologi Malaysia
Johor Bahru
Malaysia

ISSN 2192-1091   ISSN 2192-1105   (electronic)
ISBN 978-3-319-09754-1 ISBN 978-3-319-09755-8   (eBook)
DOI 10.1007/978-3-319-09755-8

Library of Congress Control Number: 2014951736

Springer Cham Heidelberg New York Dordrecht London

Printed on acid-free paper

Springer is part of Springer Science+Business Media (www.springer.com)

Preface

Tissue engineering aims to regenerate tissues and organs, which can provide biologically similar functions. Tissue engineering emerged to provide an alternative solution in order to overcome the problems of current transplantation therapy. Scaffolds play a crucial role in tissue engineering. Scaffolds function as temporary extracellular matrices for cell accommodation, proliferation and differentiation. The challenge of developing scaffolds still remains although the general requirements for scaffold are well described.

This book, *Composite Synthetic Scaffolds for Tissue Engineering and Regenerative Medicine,* makes an effort to deliver the main features and current progress of biomaterials and scaffold fabrication techniques in the area of tissue engineering and regenerative medicine. This Springer Brief consists of four chapters. Chapter 1 is anticipated to address the commonly used materials used to fabricate tissue engineering scaffolds. Chapter 2 describes the scaffold fabrication techniques. Chapter 3 focuses on fabrication and characterization of polumer and composite scaffolds by using electrospinning technique. Chapter 4 provides the production of composite scaffolds using freeze-drying technique. All these four chapters not only provide the complete summary of the current trends in fabrication of composite scaffolds but also present the new trends and directions for scaffold development for the ever expanding tissue engineering and regenerative medicine.

Malaysia, 2014

Naznin Sultana
Mohd Izzat Hassan
Mim Mim Lim

Contents

Notations

BG	Bioactive glass
cm	Centimeter
COOH	Carboxyl end group
E	Young's modulus, Ester
ECM	Extracellular matrix
FDA	Food and drug administration
g	Gram
h	Hour
HA	Hydroxyapatite
HB	Hydroxybutyrate
HV	Hydroxyvalerate
H_2O	Water
J	Joule
k, k'	Rate constant
K	Kelvin
MSC	Mesenchymal stem cells
ml	Mililiter
mm	Milimeter
MPa	Megapascal
M_{nt}	Molecular weight after in vitro degradation at time t
M_n	Number average molecular weight at time t
M_{no}	Initial number average molecular weight
NG	Nucleation and growth
PBS	Phosphate buffered saline
PCL	Poly(ε-caprolactone)
PDLA	Poly(d-lactide)
PET	Poly(ethylene terephthalate)
PGA	Poly(glycolic acid)
PHB	Poly(hydroxybutyrate)
PHBHHx	Poly(hydroxybutyrate-co-hydroxyhexanoate)

PHBV	Poly(hydroxybutyrate-co-hydroxyvalerate)
PLA	Poly(lactic acid)
PLGA	Poly(lactic acid-co-glycolic acid)
rpm	Rotational per minute
SD	Spinodal decomposition
SEM	Scanning electron microscopy
SLS	Static light scattering
t	Degradation time
T	Absolute temperature
TCP	Tricalcium phosphate
TE	Tissue engineering
TIPS	Thermally induced phase separation
W_i, W_d	Specimen weights before soaking in PBS
W_f, W_w	Specimen weights after soaking in PBS
wt	Weight
%	Percentage
σ	Tensile strength
ε	Elongation at fracture
°	Degree
°C	Degree Celsius
β-TCP	β-tricalcium phosphate
3D	Three-dimensional

Chapter 1
Scaffolding Biomaterials

Abstract The aim of tissue engineering is to develop cell, construct, and living system technologies to restore the structures and functions of damaged or degenerated tissues. Scaffolds are supporting materials used in tissue engineering applications to repair or restore damaged tissues. Biomaterials are used to fabricate scaffolds. There are different types of biomaterials including biopolymers, bioceramics and biodegradable metals. Biomaterials have to be biocompatible and nontoxic. To fabricate scaffold, appropriate biomaterial has to be chosen according to the desired characteristics and application of the scaffold. This chapter reviews different types of biomaterials for different tissue engineering applications.

Keywords Scaffolds · Biopolymers · Bioceramics · Biomaterials for scaffolds

1.1 Introduction

Biomaterial is defined as any substance or combination of substances, other than drugs, synthetic or natural in origin, which can be used for any period of time, which augments or replaces partially or totally any tissue, organ or function of the body, in order to maintain or improve the quality of life of the individual. The most common classes of biomaterials are polymers [Polycaprolactone (PCL), Polyurethane (PU), Poly(lactic acid) (PLA), Polysulfone (PS)], metals (stainless steel, titanium alloys, tantalum, gold, cobalt-chromium alloys) ceramics (alumina, carbon, titania, bioglass, hydroxyapatite (HA), zirconia) and composites (Silica/SR, HA/PE, Al_2O_3/PTFE, CF/UHMWPE). Biomaterials used for implantation should consider these properties: host response, biocompatibility, biofunctionality, functional tissue structure and pathobiology, appropriate design and manufacturability, mechanical properties, high corrosion resistance, high wear resistance, long fatigue life and adequate strength (Patel and Gohil 2012).

Mim Mim Lim and Naznin Sultana.

© The Author(s) 2015
N. Sultana et al., *Composite Synthetic Scaffolds for Tissue Engineering and Regenerative Medicine*, SpringerBriefs in Materials, DOI 10.1007/978-3-319-09755-8_1

1.2 Biomaterials

1.2.1 Biopolymers

Polymers are long chain macromolecules made up of many repeating units by covalent bonds. These repeating units are known as monomers. Most widely used biomaterials in biomedical applications are biopolymers which categorized in synthetic biopolymers and natural biopolymers. Synthetic polymers bring advantages over natural polymers in producing a wide range of degradation rates and mechanical properties. The composition of the synthetic polymers can be designed to minimize the immune response and combine the good properties together. Examples of synthetic biodegradable polymers are poly(ethylene glycol) (PEG)/ poly(ethylene oxide) (PEO), Poly(ethylene-covinylacetate) (EVA), Poly(glycolic acid) (PGA), poly(lactic acid) (PLA), polyanhydrides, polyfumarates (PF), poly- orthoesters and polycarbonates (Vail et al. 1999).

Natural polymers exhibit similar properties to soft tissues. These materials are getting from natural sources. Hence, they have to be purified to avoid foreign body response after implantation. Examples of natural polymers are gelatin, dextran, fibrin, fibronectin, agarose/alginate and hyaluronan/hyaluronic acid (Willerth and Sakiyama-Elbert 2007). Polymer is used widely because of its versatility and flexibility. It has wide range of mechanical, chemical and physical properties. It is resistance to biochemical attack, has good compatibility and light in weight. Polymers can be easily processed and shaped. Moreover, it is inert towards host tissues and available in reasonable cost. Biodegradable polymers will be selected for drug delivery system as it does not need surgery to be removed after releasing of drugs and can be excreted by body itself.

Examples of biomedical applications using biopolymers include heart valves, vascular grafts, artificial hearts, breast implants, contact lenses, intraocular lenses, components of extracorporeal oxygenators, dialyzers and plasmapheresis units, coatings for pharmaceutical tablets and capsules, sutures, adhesives, and blood substitutes, kidney, liver, pancreas, bladder, bone cement, catheters, external and internal ear repairs, cardiac assist devices, implantable pumps, joint replacements, pacemaker, encapsulations, soft-tissue replacement, artificial blood vessels, artificial skin, dentistry, drug delivery and targeting into sites of inflammation or tumors, bags for the transport of blood plasma (Patel and Gohil 2012).

1.2.2 Bioceramics

Ceramic is an organic non-metallic solid. It can be crystalline, semi-crystalline or amorphous. Bioceramics are biocompatible, bioinert, bioactive, biodegradable, soluble, resorbable, has great strength and stiffness, resistance to corrosion and wear and low in density. However, bioceramics are brittle. In order to increase the strength and elasticity of the bioceramics, some researchers blend or coat ceramic with other materials. There are three basic types of ceramics: bioinert, bioactive and bioresorbable ceramics. Bioinert ceramics are alumina (Al_2O_3), Zirconia (ZrO_2) and pyrolytic carbon. Bioactive ceramics are bioglass and glass ceramics. Bioresorbable ceramic is calcium phosphate. In bone tissue engineering, bioceramic scaffolds posses more advantages features comparing to other biomaterials (Swieszkowski et al. 2010) and proven to be more successful in small bone defects comparing to other biomaterials (Rezwan et al. 2006). By assessing whole bones in vivo, the mechanical behavior of bones can be investigated. The mechanical properties of cortical or cancellous bones are determined in vitro using standard or miniature specimens that match up to various standards originally designed for testing conventional materials such as metals and plastics (Wang 2004). It is very important to maintain the water content of bone for mechanical assessment as the behavior of bone in the "wet" condition can be significantly different from that bone in a "dry" condition (Fung 1993). Cortical bone has a range of associated properties rather than a unique set of values (Table 1.1) with respect to orientation, location and age (Wang 2004).

The mechanical behavior of bone can be explained using a simple composite model by treating bone as a nanometer-scale composite (Fig. 1.1). In bone, brittle apatite acts as a stiffening phase whereas ductile collagen provides a tough matrix. Therefore the tensile behavior of bone reveals the combinational effect of these two major constituents. A good understanding of the structure and properties of bone yields a good insight into the structural features of bones as well as provides the property range for approximating mechanical compatibility that is required of a bone analogue material for structural replacement with a stabilized bone-implant interface. It is also important to take into account that, bone can alter its properties and configuration in response to changes in mechanical demand which is unlike any engineering material.

The structure and properties of cancellous (or spongy) bone is well documented. The cancellous bone is made up of an interconnected network of rods or plates. Low density, open cells are produced by a network of rods while closed cells are produced when the rods progressively spread and flatten as the density increases.

Table 1.1 Mechanical properties of bone and current implant materials (Wang 2004)

Material	E (GPa)	σ (MPa)	ε (%)
Cortical bone	7–30	50–150	1–3
Cancellous bone	0.05–0.5	10–20	5–7
Co-Cr alloys	230	900–1,540	10–30
Stainless steel	200	540–1,000	6–70
Ti-6Al-4V	106	900	12.5
Alumina	400	450	~0.5
Hydroxyapatite	30–100	60–190	
Polyethylene	1	30	>300

E Young's modulus, σ tensile strength (flexural strength for alumina), ε elongation at fracture

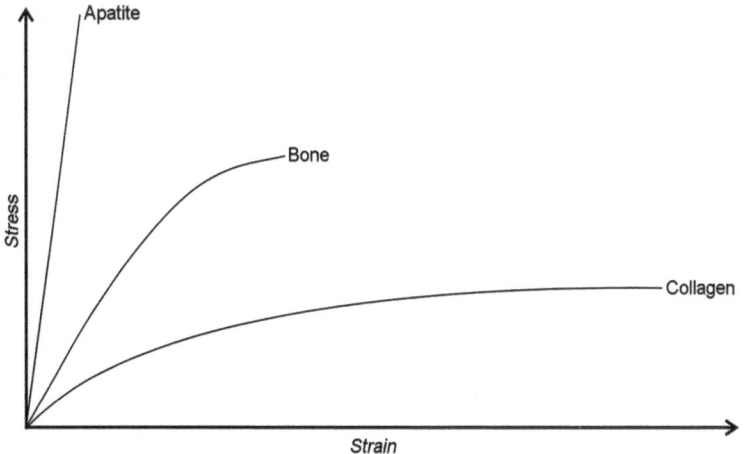

Fig. 1.1 Schematic diagram showing the mechanical behavior of apatite, collagen, and compact bone

The relative density of cancellous bone varies from 0.05 to 0.7. The compressive stress-strain curve of cancellous bone possesses the characteristics of a cellular solid. Under compression, the scaffolds exhibited linear elasticity at low stresses followed by a long plateau of cell wall collapse and then a regime of densification in which the stress rose steeply.

The linear elasticity is controlled by cell wall bending, the plateau is associated with collapse of the cells (of the "cellular structure") and when the cells have almost completely collapsed, opposing cell walls touch, with further strain compressing the solid itself, giving the final region of rapidly increasing stress. As the relative density increases, the cell walls thicken and the pore space shrinks. Increasing the relative density of the scaffold increases the compressive modulus, raises the plateau stress and reduces the strain at which densification starts.

1.2.3 Polycaprolactone (PCL)

PCL is synthetic aliphatic polyester. PCL is an implantable device material that is proved by FDA. Nowadays, PCL is often used in biomedical application because it brings advantages such as low in cost, biocompatibility, and slow biodegradability, non-toxic and has good mechanical properties (Moghe et al. 2009; Van der Schueren et al. 2011). It biocompatibility is similar to PLA and PGA. However, it has lower degradation rate as compared to PLA and PGA (Lowry et al. 1997). The degradation rate of PCL is more than 24 months (Woodruff and Hutmacher 2010) and makes it less attractive for general tissue engineering applications but it acts as a better long-term drug delivery carrier. PCL is hydrophobic and will cause limit of use in certain application (Prabhakaran et al. 2008). This property can be improved by combining PCL with a more hydrophilic polymer such as chitosan, a natural polysaccharide derived from chitin. It can provide hydrophilicity to PCL and also antimicrobial activity as well as supporting the biocompatibility of PCL (Bhattarai et al. 2009; Cooper et al. 2011; Hong and Kim 2011; Muzzarelli 2011; Prabhakaran et al. 2008; Yang et al. 2009). Besides, the porosity of scaffolds is very important in delivery system. PCL porous membranes have pore sizes on microscale. By having pores, it can control the rate of drug diffusion.

1.2.4 Poly(ortho esters) and Poly(anhydrides)

Poly(ortho ester) and polyanhydrides are biocompatibility and has degradation characteristics (Muggli et al. 1999). Previously, both of them are designed for controlled drug delivery (Burkoth et al. 2000; Hanes et al. 1998; Ibim et al. 1998) and other tissue engineering applications in recent time. The biocompatibility of poly(anhydride-co-imide) is equal to PLGA (Ibim et al. 1998). Its scaffold can be used in orthopaedic surgery and also weight-bearing applications (Burkoth et al. 2000; Liu et al. 2007).

1.2.5 Poly(glycolic acid), Poly(lactic acid) and Their Copolymers

Scaffolds Poly(glycolic acid) (PGA), poly(lactic acid) (PLA), and their copolymers are a family of linear aliphatic polyesters. They are synthetic degradable polymers which are approved by US Food and Drug Administration (FDA) for certain human clinical applications. They are also widely used in medicine and tissue engineering applications. PGA is linear aliphatic polyester that has the simplest structure as shown in figure below. PGA is highly crystalline, low solubility in organic solvents and has high melting point. PGA was investigated for the first synthetic absorbable suture (Frazza and Schmitt 1971). However, the sutures lose their mechanical strength due to hydrophilicity of PGA (Reed and Gilding 1981). PLA is more hydrophobic and more soluble in organic solvent than PGA as it has an extra methyl group in its repeating unit. Duration for PLA to lose its mechanical integrity is longer than PGA. In order to achieve intermediate degradation rate between PGA and PLA, their copolymer, poly(lactic acid-co-glycolic acid) (PLGA) was synthesized by using various ratios of lactic and glycolic acid (Sultana and Wang 2012). There are some morphological changes for PGA-PLA copolymers. Crystallinity is lost. PGA-PLA copolymers are investigated for drug delivery applications and this effort has been comprehensively reviewed by Lewis (Lanza et al. 2007).

1.2.6 Chitosan

Chitosan is a natural polymer which formed at about 50 % degree of deacetylation of chitin. Chitosan is soluble in aqueous acidic solution. Chitosan is very unique of its characteristic which is the only pseudonatural cationic polymer. Chitosan is a semicrystalline polymer in solid state. Chitosan is much easier to be process than chitin but it has lower stability and is hydrophilic as well as pH sensitivity (Rinaudo 2006). Chitosan can be processed into different forms including scaffolds, bandages, micro and nanoparticles, nanofibers, micro and nanogel, film, beads, hydrogel, membranes and sponges (Anitha et al. 2014). Examples of the derivatives of chitosan are N-methylene phosphonic chitosans, O- and N-carboxymethylchitosans, chitosan 6-O-sulfate, trimethylchitosan ammonium, carbohydrate branched chitosans, chitosan-grafted copolymers, alkylated chitosans, and cyclodextrin-linked chitosans (Rinaudo 2006). Chitosan is used widely in various applications including agriculture, water and waste treatment, food and beverages, cosmetics and toiletries, biopharmaceutics, biomedical applications and drug delivery applications (Anitha et al. 2014). Tiyaboonchai concluded that chitosan nanoparticles are prospective drug delivery carriers as they are safe, biocompatible and biodegradable. Moreover, chitosan has the ideal property for drug delivery carrier which is water soluble. It is also available in wide range of molecular weights and can be chemically modified easily. It also shows absorption enhancing effect. Tiyaboonchai also concluded that chitosan nanoparticles is a good adjuvant for vaccine (Tiyaboonchai 2013). Stephen-Haynes and his researchers had produced a product named KytoCel for wound healing and promote skin regeneration by using chitosan. KytoCel is not toxic, not causing irritation, non-immunogenicity, degradable and has good biocompatibility (Stephen-Haynes et al. 2014).

1.2.7 Gelatin

In fact, Gelatin is a natural polymer in the mixture of peptides and proteins. It is derived from collagen by denaturing or partially hydrolysis. Gelatin can be easily obtained from animal skin, white connective tissue or muscle, tendon, ligament and bones. Gelatin is brittle and faintly yellow in colour. Gelatin can be dissolved in aqueous solutions of polyhydric alcohol such as glycerol and also organic solvents which are highly polar and has hydrogen bonding such as acetic acid (Finch and Jobling 1977). Gelatin is more soluble in hot than cold water and insoluble in less polar organic solvents. Gelatin has amphoteric properties where it can act as acid or base. Gelatin is widely used in food, pharmaceutical, photographic industries and diverse technical uses. Nowadays, gelatin is also used as a material for fabricating scaffolds. Gelatin as a natural polymer brings advantages in biomedical applications. It is biocompatible, biodegradable, and low in cost. Ki, Zhang and their researchers had successfully fabricated uniform and very fine gelatin nanofibers using electrospinning technique (Ki et al. 2005; Zhang et al. 2005).

1.2.8 Biodegradable Metals

Metals are mostly used for load bearing implants. Biomedical metals that are widely used are stainless steel, cobalt-chromium-molybdenum alloy, titanium and titanium alloys. Metals are selected as good implantable material because of their excellent electrical and thermal conductivity, appropriate mechanical properties, corrosion resistance, biocompatibility and the cost is reasonable (Patel and Gohil 2012). However, implantable biomaterials are only temporary needed for supporting healing of injuries. It is better if these biomaterials can be fully degraded in human body environment after the healing period and do not cause toxicity to the body during the corrosion process (Hermawan 2012; Witte and Eliezer 2012).

Nowadays, researchers are investigating metals that have controlled corrosion properties for implantable devices which are also called biodegradable metals. The typical way of degradation of metals is through corrosion. Corrosion of implanted metals is due to many reasons including local pH for magnesium and local oxygen concentration for iron (Witte and Eliezer 2012). Biodegradable metals can be classified as pure biodegradable metals (magnesium-based and iron-based), bio-degradable alloys and biodegradable metal matrix composites. According to Zheng et al. (2014) biodegradable metals are bioactive materials and its research should move towards to third generation biomedical materials with various controlled capabilities to benefit local tissue reconstruction.

Iron is an essential trace element in the body and only Fe^{2+} can be absorbed by intestine. Fe^{3+} has to be reduced to Fe^{2+} before it can be absorbed. The most important thing that needs to take note for biodegradable iron implants is the patient should not has iron-related disease and the local storage capacity and the amount of iron released from the implant. The amount of iron released should not exceed the local storage capacity (Witte and Eliezer 2012). Iron based biodegradable metal has similar mechanical properties as stainless steel. However, it shows slow degradation rate and not MRI compatible. Although iron has slow degradation rate, it does not cause toxicity. Some new fabrication methods emerged to increase the degradation rate and improve MRI compatibility such as casting, powder metallurgy, electro-forming and inkjet 3D-printing (Zheng et al. 2014).

Biodegradable iron implants has been developed mainly for cardiovascular stent applications. The first biodegradable iron stents was implanted into descending aortas of New Zealand white rabbits by Peuster et al. (2001). In the research on coronary arteries of young pigs, the performance of iron stents was similar to Co-Cr stents (Waksman et al. 2008). Moreover, the released ions by the corrosion of iron stent reduce the cell proliferation of vascular smooth muscle cells (Mueller et al. 2006). Hence, the corrosion rate of iron stent designed is very important to avoid toxicity to cells.

1.2.9 Biodegradable Polymer Blends

Blending of biodegradable polymers can improve the performance and reduce expense. Blends of natural polymers with synthetic polymers can be used to improve the degradation properties. PLA is biodegradable and non-toxic to the human body. PLA possesses high mechanical performance similar to some commercial polymers such as polyethylene and poly(ethylene terephthalate) (PET). Because of good biodegradability and very low toxicity, PLA based materials have been widely used for biomedical and pharmaceutical applications such as fixation of fractured bone and matrices for drug delivery systems. Depending on their applications, the physical properties and biodegradation behavior, biodegradation kinetics of PLA can be modified by blending. Blends of PLA with poly(d-lactide) (PDLA) can be used to prepare novel hydrogels and microspheres for biomedical applications. It was reported that drug delivery system particles were fabricated from l-configured peptides such as insulin with PDLA, PDLA-b-PEG, PDLA-b-PEG/PDLA, PLLA/PDLA or PLLA-b-PEG/PDLA-b-PEG.

As biodegradable PHB is very brittle and prone to thermal degradation, in order to improve its mechanical properties and processability, blending with another polymer can be done. There are several reports on the blending of PHB with other biodegradable polymers including poly(l-lactic acid-co-ethylene glycol-co-adipic acid), PCL, PHBV.

Blends of PLLA with two kinds of PHB with different molecular weights were prepared by Park and co-workers by the solvent casting method. It was reported from DSC analysis that the system was immiscible over the entire composition range. It was also found that the mechanical properties of all the samples were improved. A good interfacial adhesion between two polymers and the reinforcing role of PLLA components led to enhanced mechanical properties to the PLLA content.

It was reported that the presence of a second component in the blend with PHBV, whatever its chemical nature, is sufficient to perturb the crystallization behavior of highly crystalline PHBV and enhance hydrolytic degradation. The introduction of polar carboxylic groups in side-chains led to an increase in the degradation rate as carboxylic groups promote water penetration into the polymer.

1.2.10 Composites

Composite materials are solids containing two or more distinct constituent materials or phases on a scale larger than the atomic. Composite materials usually exhibit controllable mechanical properties such as stiffness, strength, toughness etc. Ceramic/polymer composites exhibit the best characteristics of each constituent, i.e. the toughness of polymer and stiffness of ceramic. Artificial ceramic/polymer composites are usually produced as analogue biomaterials for bone substitute as natural bone is a collagen/apatite composite. Synthetic polymer/naturally derived polymer composite scaffold composed of PLGA and collagen was produced and it was reported that the biological activities of the composite scaffolds could be helpful in bone tissue regeneration and that the composite scaffolds were able to promote cellular interactions. Similarly, a blend scaffold of PCL/PLG/hydroxyapatite was fabricated for the applications in bone tissue regeneration and it was reported that the composite scaffolds are able to promote cellular interactions. As no single material has been shown to be able to meet the requirements for bone tissue engineering, composites seem to be the most promising way in future.

References

Anitha, A., Sowmya, S., Kumar, P. T., Deepthi, S., Chennazhi, K. P., Ehrlich, H., et al. (2014). Chitin and chitosan in selected biomedical applications. *Progress in Polymer Science, 39,* 1644–1667.

Bhattarai, N., Li, Z., Gunn, J., Leung, M., Cooper, A., Edmondson, D., et al. (2009). Natural-synthetic polyblend nanofibers for biomedical applications. *Advanced Materials, 21,* 2792–2797.

Burkoth, A. K., Burdick, J., & Anseth, K. S. (2000). Surface and bulk modifications to photocrosslinked polyanhydrides to control degradation behavior. *Journal of Biomedical Materials Research, 51,* 352–359.

Cooper, A., Bhattarai, N., & Zhang, M. (2011). Fabrication and cellular compatibility of aligned chitosan–PCL fibers for nerve tissue regeneration. *Carbohydrate Polymers, 85,* 149–156.

Finch, C. A., & Jobling, A. (1977). The physical properties of gelatin. In *The science and technology of gelatin.* London: Academic Press.

Frazza, E., & Schmitt, E. (1971). A new absorbable suture. *Journal of Biomedical Materials Research, 5,* 43–58.

Hanes, J., Chiba, M., & Langer, R. (1998). Degradation of porous poly(anhydride-co-imide) microspheres and implications for controlled macromolecule delivery. *Biomaterials, 19,* 163–172.

Hermawan, H. (2012). *Biodegradable metals: From concept to applications.* New York: Springer.

Hong, S., & Kim, G. (2011). Fabrication of electrospun polycaprolactone biocomposites reinforced with chitosan for the proliferation of mesenchymal stem cells. *Carbohydrate Polymers, 83,* 940–946.

Ibim, S. M., Uhrich, K. E., Bronson, R., El-Amin, S. F., Langer, R. S., & Laurencin, C. T. (1998). Poly(anhydride-co-imides): In vivo biocompatibility in a rat model. *Biomaterials, 19,* 941–951.

Ki, C. S., Baek, D. H., Gang, K. D., Lee, K. H., Um, I. C., & Park, Y. H. (2005). Characterization of gelatin nanofiber prepared from gelatin-formic acid solution. *Polymer, 46,* 5094–5102.

Lanza, R., Langer, R., & Vacanti, J. (2007). Principles of tissue engineering. London: Academic Press.

Liu, C., Xia, Z., & Czernuszka, J. (2007). Design and development of three-dimensional scaffolds for tissue engineering. *Chemical Engineering Research and Design, 85,* 1051–1064.

Lowry, K., Hamson, K., Bear, L., Peng, Y., Calaluce, R., Evans, M., et al. (1997). Polycaprolactone/glass bioabsorbable implant in a rabbit humerus fracture model. *Journal of Biomedical Materials Research, 36,* 536–541.

Moghe, A., Hufenus, R., Hudson, S., & Gupta, B. (2009). Effect of the addition of a fugitive salt on electrospinnability of poly(ε-caprolactone). *Polymer, 50,* 3311–3318.

Mueller, P. P., May, T., Perz, A., Hauser, H., & Peuster, M. (2006). Control of smooth muscle cell proliferation by ferrous iron. *Biomaterials, 27,* 2193–2200.

Muggli, D. S., Burkoth, A. K., & Anseth, K. S. (1999). Crosslinked polyanhydrides for use in orthopedic applications: Degradation behavior and mechanics. *Journal of Biomedical Materials Research, 46*, 271–278.

Muzzarelli, R. A. (2011). Biomedical exploitation of chitin and chitosan via mechano-chemical disassembly, electrospinning, dissolution in imidazolium ionic liquids, and supercritical drying. *Marine Drugs, 9*, 1510–1533.

Patel, N. R., & Gohil, P. P. (2012). A Review on biomaterials: Scope, applications & human anatomy significance. *International Journal of Emerging Technology and Advanced Engineering, 2*, 91–101.

Peuster, M., Wohlsein, P., Brügmann, M., Ehlerding, M., Seidler, K., Fink, C., et al. (2001). A novel approach to temporary stenting: Degradable cardiovascular stents produced from corrodible metal—Results 6–18 months after implantation into New Zealand white rabbits. *Heart, 86*, 563–569.

Prabhakaran, M. P., Venugopal, J. R., Chyan, T. T., Hai, L. B., Chan, C. K., Lim, A. Y., et al. (2008). Electrospun biocomposite nanofibrous scaffolds for neural tissue engineering. *Tissue Engineering Part A, 14*, 1787–1797.

Reed, A., & Gilding, D. (1981). Biodegradable polymers for use in surgery—poly (glycolic)/poly (lactic acid) homo and copolymers: 2. In vitro degradation. *Polymer, 22*, 494–498.

Rezwan, K., Chen, Q. Z., Blaker, J. J., & Boccaccini, A. R. (2006). Biodegradable and bioactive porous polymer/inorganic composite scaffolds for bone tissue engineering. *Biomaterials, 27*, 3413–3431.

Rinaudo, M. (2006). Chitin and chitosan: Properties and applications. *Progress in Polymer Science, 31*, 603–632.

Stephen-Haynes, J., Gibson, E., & Greenwood, M. (2014). Chitosan: A natural solution for wound healing. *Journal of Community Nursing, 28*, 48–53.

Sultana, N., & Wang, M. (2012). PHBV/PLLA-based composite scaffolds fabricated using an emulsion freezing/freeze-drying technique for bone tissue engineering: Surface modification and in vitro biological evaluation. *Biofabrication, 4*, 015003.

Swieszkowski, W., Jaegermann, Z., Hutmacher, D. W., & Kurzydlowski, K. J. (2010). Ceramic materials for bone tissue replacement and regeneration. In D. Jiang, Y. Zeng, M. Singh, & J. Heinrich (Eds.), *Ceramic materials and components for energy and environmental applications* (pp. 525–530). Hoboken, NJ, USA: John Wiley & Sons, Inc. doi: 10.1002/9780470640845.ch74.

Tiyaboonchai, W. (2013). Chitosan nanoparticles: A promising system for drug delivery. *Naresuan University Journal, 11*, 51–66.

Vail, N., Swain, L., Fox, W., Aufdlemorte, T., Lee, G., & Barlow, J. (1999). Materials for biomedical applications. *Materials and Design, 20*, 123–132.

Van der Schueren, L., de Schoenmaker, B., Kalaoglu, Ö. I., & de Clerck, K. (2011). An alternative solvent system for the steady state electrospinning of polycaprolactone. *European Polymer Journal, 47*, 1256–1263.

Waksman, R., Pakala, R., Baffour, R., Seabron, R., Hellinga, D., & Tio, F. O. (2008). Short-term effects of biocorrodible iron stents in porcine coronary arteries. *Journal of Interventional Cardiology, 21*, 15–20.

Willerth, S. M., & Sakiyama-Elbert, S. E. (2007). Approaches to neural tissue engineering using scaffolds for drug delivery. *Advanced Drug Delivery Reviews, 59*, 325–338.

Witte, F., & Eliezer, A. (2012). Biodegradable metals. In *Degradation of implant materials*. New York: Springer.

Woodruff, M. A., & Hutmacher, D. W. (2010). The return of a forgotten polymer—Polycaprolactone in the 21st century. *Progress in Polymer Science, 35*, 1217–1256.

Yang, X., Chen, X., & Wang, H. (2009). Acceleration of osteogenic differentiation of preosteoblastic cells by chitosan containing nanofibrous scaffolds. *Biomacromolecules, 10*, 2772–2778.

Zhang, Y., Ouyang, H., Lim, C. T., Ramakrishna, S., & Huang, Z. M. (2005). Electrospinning of gelatin fibers and gelatin/PCL composite fibrous scaffolds. *Journal of Biomedical Materials Research Part B: Applied Biomaterials, 72,* 156–165.

Zheng, Y. F., Gu, X. N., & Witte, F. (2014). Biodegradable metals. *Materials Science and Engineering: R: Reports, 77,* 1–34.

Chapter 2
Scaffold Fabrication Protocols

Abstract Development of scaffolds in tissue engineering applications is growing in a fast pace. Scaffolds play a pivotal role in scaffold-based tissue engineering. The scaffolds must possess some important characteristics. Scaffolds should be biocompatible, should have appropriate porosity and porous microstructure and proper surface chemistry to allow cell attachment, proliferation and differentiation. Scaffolds should possess adequate mechanical properties and controlled biodegradability. There are many techniques available to fabricate scaffolds including freeze drying, electrospinning and rapid prototyping. Some of these techniques have gained much attention due to their versatility. This chapter points up the protocols for the fabrication and characterization of appropriate scaffolds for tissue engineering using biopolymers and composite biomaterials.

Keywords Fabrication · Freeze-drying · Electrospinning · Rapid prototyping

2.1 Techniques of Producing Scaffolds

There are many techniques used in making scaffolds and divided into two categories, non-designed manufacturing techniques and designed manufacturing techniques. Non-designed manufacturing techniques include freeze drying or emulsion freezing (Whang et al. 1995), solvent casting or particulate leaching, phase separation (Zhao et al. 2002; Liu and Ma 2004; Ma 2004), gas foaming or high pressure processing, melt moulding, electrospinning and combination of these techniques. Designed manufacturing technique includes rapid prototyping of solid free-form technologies.

Mim Mim Lim and Naznin Sultana.

© The Author(s) 2015
N. Sultana et al., *Composite Synthetic Scaffolds for Tissue Engineering and Regenerative Medicine*, SpringerBriefs in Materials,
DOI 10.1007/978-3-319-09755-8_2

2.1.1 Electrospinning

To process solutions or melts of polymers into continuous fibers into continuous fibers with diameters ranging from nanometers to submicrometers, electrospinning is a highly versatile method. In this process, a polymer solution is contained in a syringe with a metal capillary connected to a high voltage power supply as an electrode. Subjected to an electric field, the polymer jet travels in the electric field which becomes thin fibers and deposits onto a conductive collector. The ultrafine fibers in the electrospun mats have high surface-to-volume ratios and porosity and mimic the natural extracellular matrix of body tissues. In order to regenerate various tissues including skin, blood vessel, cartilage, bone, muscle, ligament and nerve, electrospun fibers have been employed.

2.1.2 Solvent Casting and Particulate Leaching

To overcome the drawbacks associated with the fiber bonding technique, a solvent-casting and particulate-leaching technique was developed. Porous constructs of synthetic biodegradable polymers could be prepared with specific porosity, surface to volume ratio, pore size and crystallinity for different applications by appropriate thermal treatment. Though the technique was valid for PLLA and PLGA scaffolds, this could be applied to any other polymer that was soluble in a solvent such as chloroform or methylene chloride.

2.1.3 Polymer Phase Separation

Thermally induced phase separation (TIPS) of polymer solution was reported to be used in the field of drug delivery and to fabricate microspheres in order to incorporate biological and pharmaceutical agents (Ma 2004). In order to fabricate tissue engineering scaffolds, this process has become much popular. By altering the types of polymer and solvent, polymer concentration and phase separation temperature, different types of porous scaffolds with micro and macro-structured foams can be produced. In order to control pore morphology on a micrometer to nanometer level, TIPS process can be utilized (Ma 2004).

Depending on the thermodynamics and kinetic behaviour of the polymer solution under certain conditions, TIPS can be a complicated process. It was defined that if a system where the solvent crystallization temperature (freezing point) is higher than the liquid-liquid phase separation temperature, the system can separate by lowering the temperature, the process is called solid-liquid phase separation. After the removal of the solvent, the remaining pores have morphologies similar to solvent crystallite geometries. On the other hand, when the solvent crystallization temperature is much lower than the phase separation temperature, if the temperature of polymer solution is decreased, a liquid-liquid phase separation takes place.

2.1.4 Rapid Prototyping

Rapid prototyping of solid free-form is a technique designed manufacturing technique that fabricates three-dimensional scaffolds with fully interconnected porous network (Lam et al. 2002). It includes three-dimensional printing, laser sintering and stereolithography. These techniques require a computer model of the desired scaffold architecture from computer-assisted design (CAD) or computed tomography (CT). The advantages of these techniques are their capability to fabricate scaffolds with complex architectures in micron scale μm (Lanza et al. 2007). There are some interesting and unique challenges in scaffold design (Lanza et al. 2007):

- Almost all bone defects are irregularly shaped; any proposed scaffold processing technique must be sufficiently versatile to allow the formation of porous polymer-based materials with irregular three-dimensional shape.
- The scaffold must have high strength to replace the structural function of bone temporarily until it is regenerated.

For many orthopedic applications, poly(α-hydroxyesters) were used in a solid form, but the compressive strength of foam scaffolds constructed of these materials rapidly decreased with increasing porosity. In order to formulate polymer/ceramic composites, an alternative method was proposed using a novel phase transition technique. Hydroxyapatite powder was added to a PLGA/dioxane solution according to this process. The mixture was then frozen for several hours to induce phase separation and then freeze dried to sublimate the solvent. The composite foams thus produced, exhibited interconnected irregular pore morphology with a polymer/hydroxyapatite skeleton. The compressive strength of these foams was significantly higher than that of foams made from pure PLGA. The porosity, pore size, and pore structure could be controlled by changing the polymer concentration, hydroxyapatite amount, solvent type and phase separation temperature. Composite foams with porosity of up to 95 % and pore size in the range of 30–100 μm were fabricated with this method.

2.1.5 Melt Molding

Another alternative method of constructing three-dimensional scaffolds is melt molding. By using this technique, PLGA scaffolds were produced by leaching PLGA/gelatin microsphere composites. According to this method, a fine PLGA powder was mixed with previously sieved gelatin microspheres and poured into a Teflon mold, which was then heated above the glass transition temperature of the polymer. The PLGA/gelatin microsphere composite was then removed from the mold and placed in distilled-deionized water. The water soluble gelatin was leached out leaving a porous PLGA scaffold with the geometry identical to the shape of the mold. It was possible to construct PLGA scaffolds of any shape simply by changing the mold geometry by using this method. The porosity could be controlled by varying the amount of gelatin used to construct the composite material and the pore size of the scaffold could also be altered independently of the porosity by using different diameters of microspheres. Another advantage of this method was that it does not utilize organic solvents and is carried out at relatively low temperatures. For this reason, it had the potential for the incorporation and controlled delivery of bioactive molecules. This scaffold manufacturing technique could also be applied to other polymers such as PLLA and PGA. Many of the scaffold preparation design criteria were satisfied by this technique and offer an extremely versatile means of scaffold preparation. Alternative leachable components such as salt or other polymer microspheres could also be used other than gelatin microspheres.

2.1.6 Gas Foaming

In order to eliminate the need for organic solvent in the pore-making process, a new technique involving gas as a porogen was introduced. The process started with the formation of solid discs of PGA, PLLA or PLGA by using compression molding with a heated mold. The discs were placed in a chamber and exposed to high pressure CO_2 (5.5 MPa) for 3 days. At this time, pressure was rapidly decreased to atmospheric pressure. Up to 90 % porosities and pore sizes of up to 100 μm could be obtained using this technique. But the disadvantage was pores are largely unconnected, especially on the surface of the scaffold. Although the fabrication method required no leaching step and used no harsh chemical solvents, but the high temperature which was involved in the disc formation, prohibited the incorporation of cells or bioactive molecules. Also the unconnected pore structure made cell seeding and migration within the scaffold difficult. In order to produce an open pore morphology using this technique, both gas foaming and particulate leaching technique were developed. According to this, salt particles and PLGA pellets were mixed together and compressed to scaffold solid disks which were saturated with high pressure gas and the pressure was subsequently reduced. The salt particles were removed then by leaching. This combination guided to a porous polymer matrix with an open, interconnected morphology without the use of any organic solvents. This technique might have widespread use in cell transplantation applications of many types of cells, including hepatocytes, chondrocytes, and osteoblasts.

2.2 Electrospinning Protocol

The technique of using electrostatic forces to produce synthetic fiber has emerged since 100 years ago (Sill and von Recum 2008). In the early stage, electrospinning has not emerged as a feasible technique to produce small diameter of polymer fiber because of some technical difficulties. In the year 1934, Formhals patented a process and apparatus to spin fiber by using electrical charges (Formhals 1934). The apparatus use a moveable threat-collecting device to collect fiber. Aligned fibers can be collected by this apparatus. Formhals use acetone/alcohol solution as solvent to produce cellulose acetate fibers was a success by using this apparatus. However, it forms loose web structure due to the short distance between spinning and the collector. The solvent was not fully evaporated. Besides, incomplete evaporation of the solvent will also lead to the fiber stick on the collector making removal problematic.

In the second patent, Formhals detailed the apparatus by setting a bigger distance between the spinning and the collector in order to solve the problem faced which is incomplete evaporation of the solvent (Anton 1939). Formhals described the use of multiple nozzles. In 1940, Formhals patented a new process. In the process, composite fiber is formed by directly electrospun polymer solution onto a moving base thread (Formhals 1940). Taylor published his work in 1969 about Taylor cone (Taylor 1969). He found that when surface tension is balanced with electrostatic forces, the pendant droplet will form into a cone which is named by his name, Taylor cone. The fiber jet is emitted from the apex of the Taylor cone and this generates a smaller diameter of fiber produced which is smaller than the diameter of capillary tip. Taylor also determined the angle that is needed to balance the surface tension with electrostatic forces which is 49.3° with respect to the axis of the cone at cone apex or 98.6° of cone angle.

Besides, there are some parameters that can affect the structural properties of the electrospun fibers produced. The investigate of varying these parameters such as solution and processing parameters for example solution viscosity, applied voltage and flow rate is done by Baumgarten in 1971. The solution used by Baumgarten is polyacrylonitrile/dimethylformamide (PAN/DMF). In his investigation, he found out that higher viscosities will give larger fiber diameter. For electrical field, while continuously increasing the applied voltage, the fiber diameter will decrease initially until a minimum and then increase.

Approximately a decade after Baumgarten's work, there are other works began on electrospinning polymer melts. Larrondo and Mandley electrospun fibers from polyethylene and polypropylene melts and it was a success (Baumgarten 1971; Larrondo and St John Manley 1981). The difference between electrospun fiber produced from solution and melt is electrospun fiber from a melt has larger diameters. Larrondo and Mandley had discovered that melt temperature can inversely influence the fiber diameter.

Investigation of potential applications of electrospun fiber began after these findings. One of the applications is in tissue engineering field. Annis and Bornat published their work on the application of electrospun polyurethane mats as vascular prosthesis in 1978 (Larrondo and St John Manley 1981). In 1985, long-term in vivo performance of electrospun arterial prosthesis was examined by Fisher and Annis (Annis et al. 1978; Fisher et al. 1984).

2.2.1 Electrospinning Process and Principle

There are three major components in electrospinning, high voltage power supply, spinneret/needle and collector which are normally metal screen and are grounded. DC voltage or AC voltage can be used. Figure 2.1 shows the schematic illustration of electrospinning. Electrospinning is a versatile and very simple process. It uses electrostatic force to produce fibers by electrospin with polymer solution or polymer melt. The diameter of the polymer fibers produced ranging from a few nanometers to micrometers.

Syringe pump is used to force the polymer solution through needle at a controlled rate (mL/h). A high voltage supply is connected to the needle to inject charge into polymer solution. Electrostatic force which is the repulsion of similar charges will be produced. When electrostatic force is balanced with surface tension, Taylor cone will be formed. By increasing the electric field, fiber jet will be ejected from the apex of Taylor cone when electrostatic force is larger than surface tension.

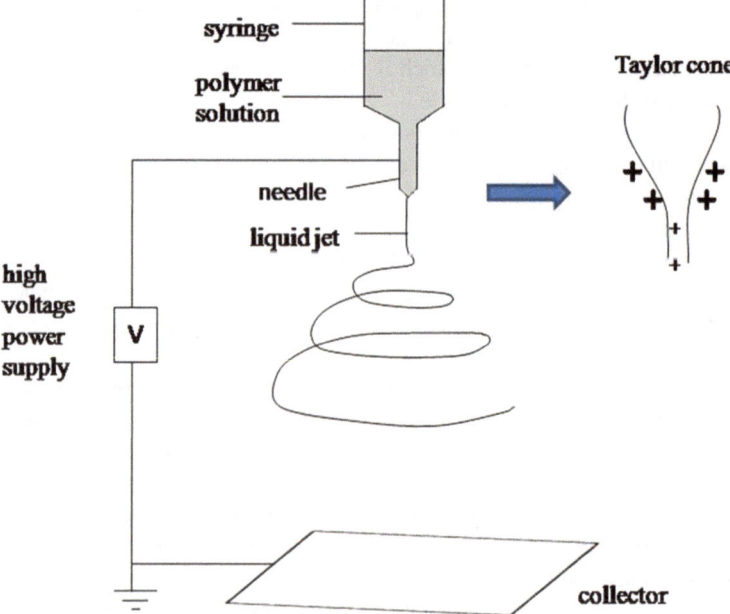

Fig. 2.1 Schematic illustration of electrospinning

During the ejection and elongation process, solvent is evaporated. The elongation and evaporation of the solvent reduces the diameter of the fiber from micrometer to nanometer. The electrified jet will undergo stretching and whipping process, long and thin thread will be formed. Lastly, the polymer fiber produced is attracted by grounded collector which is a metal screen (Doshi and Reneker 1995; Sill and von Recum 2008; Li and Xia 2004; Yu et al. 2009; Reneker and Chun 1996).

The polymer fibers produced by electrospinning process is called electrospun membrane or nanofiber. It has high porosity and similar to natural extracellular matrix (ECM). These fibers can enhance cell attachment, mass transfer and drug loading properties (Sill and von Recum 2008).

2.2.2 Nozzle Configuration

Nozzle Configuration can influence the electrospun fiber produced. There are a few nozzle configurations such as single, single with emulsion, side-by-side, and coaxial nozzles. Choosing of suitable nozzle configurations depends on the application and the product that want to be developed (Sill and von Recum 2008).

Single nozzle configuration technique is the most simple and common technique where charged polymer solution flows through single capillary. For spinning single polymer solution and polymers that is soluble in a common solvent, this configuration technique is suitable (Stitzel et al. 2006; Tan et al. 2005). For an example, poly(ethylene oxide) (PEO) was electrospun by this technique (Doshi and Reneker 1995).

Side-by-side nozzle configuration technique is used for the while desired polymer is not soluble in common solvent. In this configuration, two capillaries that located side-by-side contain two different polymer solutions. The ability of side-by-side configuration to form a single fiber jet is influenced by solution conductivity. Both polymer solutions have to have similar conductivities in order to form a single Taylor cone. Once the single Taylor cone is produced, by increasing the applied voltage, fiber jet that containing both polymers will be developed but the amount of every polymer will vary along the fiber. For an example, using of poly(vinyl chloride)/segmented polyurethane and poly(vinyl chlorine)/poly(vinylidiene fluoride) to electrospin biocomponent systems by this configuration (Gupta and Wilkes 2003).

Another potential configuration technique is coaxial configuration. In this configuration, two different polymers flow through two capillaries which are coaxial. A smaller capillary is inside the larger capillary. While applying an external voltage, Taylor cone is formed at the tip of the capillary. It will lead to the formation of the fiber where one polymer fiber is encapsulated by the other. This is known as core-shell morphology. The core polymer solution can either be or not be electrospinnable but the shell polymer solution has to be electrospinnable. It has the advantage that this technique is able to protect easily denatured biological agents effectively. It also can wrap all substances in the core regardless the interaction between the core and the shell. It is a good technique to electrospin fibers containing drugs, genes, protein and growth factors (Huang et al. 2006; Li and Xia 2004; Loscertales et al. 2002). There are some researches on coaxial configuration technique in drug delivery application that successfully mitigated the initial burst release. The drugs are released in a controlled way and have longer sustained release. Zhang et al. encapsulated model protein, fluorescein isothiocyanate conjugated bovine serum albumin with poly(ethylene glycol) (PEG) in poly(e-caprolactone) (PCL) fibers (Huang et al. 2006). Huang et al. used PCL as the shell and medically pure drugs, Resveratrol and Gentamycin Sulfate as cores (Zhang et al. 2006).

2.2.3 Solution Versus Melt Electrospinning

Two forms of polymers can be used for electrospinning, solution and melt. A polymer can dissolve in suitable solvent or directly electrospinning from its melt. Both forms will produce different size of fibers. For solution form of polymer, the fiber produced has a greater range of size, microns to nanometer. However, for melt form of polymer, the fiber sizes produced are only in micron size (Larrondo and St. John Manley 1981).

Furthermore, there are some advantages and disadvantages of both solution and melt form of polymer. Melt electrospinning does not need any solvent but it has to be kept at high temperatures to be electrospun. However, solution electrospinning can be done at room temperature. For polymer in the form of solution where the solvent is very volatile, it will reduce the mass flow velocity (Reneker and Chun 1996). High temperature used for melt electrospinning will preclude the use of fiber in some applications such as tissue engineering field and also drug delivery system. Dalton et al. has discovered that the optimum melt temperature ranged between 60 and 90° for melt electrospinning of copolymers of poly(ethylene glycol) PEG and PCL (Dalton et al. 2006). Melt electrospinning will eliminate the problem of insufficient evaporation of the solvent during the ejection of fiber jet. However, the temperature needs to be able to be cooled before reaching the collector.

2.3 Parameters in Electrospinning

In recent years, the production of electrospun fibers has gained importance due to its versatility. However, there are some parameters that influence the structure and morphology of the electrospun fibers produced. The parameters are divided into two categories, processing parameters and solution parameters.

2.3.1 Processing Parameters

2.3.1.1 Applied Voltage

Applied voltage is the high voltage that applied to the needle. Increasing of applied voltage will lead to the formation of Taylor cone. While increasing the voltage again, it will result in fiber jet being ejected from the apex of the Taylor cone.

V. Sencadas examined the effect of applied voltage on electrospinning of poly (vinylidene fluoride–trifluoroethylene) (PVDF–TrFE) at 15/85 (15 % PVDF–TrFE + 85 % solvent blend). The tip to collector distance is 20 cm, needle inner diameter is 0.5 mm, flow rate 0.5 ml/h and applied voltages of 15 and 35 kV. Travelling distance, needle inner diameter and flow rate are kept constant. In this experiment, there is small variations occurred. However, there is a trend that by increasing the applied voltage, fiber diameter is decreasing. It is due to variations of mass flow and jet dynamics (Sencadas et al. 2012).

2.3.1.2 Flow Rate

To investigate the influences of solution flow rate on fiber size and distribution, applied voltage and distance between capillary tip and collector need to be kept constant. Flow rate of solution brings impact on fiber size. It can also affect fiber porosity as well as fiber shape. Megelski et al. (2002) investigated polystyrene/ tetrahydrofuran (THF) solution on this parameter. The result showed that increasing flow rate will increase fiber diameter and pore size. However, at high flow rate, bead defects will form due to insufficient evaporation of solvent. The fiber is not completely dry before reaching collector. The increase of flow rate increases the volume from the needle tip. It will lead to increase of evaporation time of the solvent and larger polymer crystallization time. Hence, a larger diameter of fiber and broader size distribution will be produced.

2.3.1.3 Capillary Tip to Collector Distance

During fiber jet ejection, evaporation of the solvent occurs. Fiber need to be completely dry before reaching the collector to avoid forming of bead defects where beads will form on fibers. Hence, the distance between capillary tip and grounded collector plays an important role to achieve a good fiber structure. A shorter fiber travelling distance will decrease the solvent evaporation time, results in insufficient evaporation. Fiber diameter decreases with increasing distance between capillary tip and collector. Megelski et al. showed the formation of beaded electrospun polystyrene fibers due to shortening the travelling distance which attributed to inadequate drying of polymer fiber prior reaching the collector (Sill and von Recum 2008).

2.3.2 Solution Parameters

2.3.2.1 Polymer Concentration

Surface tension and viscosity of solution depends on polymer concentration. Polymer concentration will affect the spinnability of the solution. The solution needs to have high polymer concentration for chain entanglements to occur but not too concentrated. If the solution is too concentrated, the fibers could not be formed because the viscosity of the solution is too high. However, if the solution is too dilute, the polymer fibers will be broken up into droplets before reaching the collector. It is because of the effects of surface tension. Moreover, within the optimal range of polymer concentration, increasing of polymer concentration will lead to increasing of fiber diameter (Sill and von Recum 2008).

2.3.2.2 Solvent Volatility

Selection of solvent is a very important task. A volatile solvent must be used in order for sufficient solvent evaporation to occur. Decreasing solvent volatility will increase the pore size with decreased pore depth. The pore density will be decreased. Megelski et al. (2002) investigated the structural polystyrene fibers properties by varying the solvent. Solvents used are dimethylformamide (DMF) and tetrahydrofuran (THF). THF is more volatile than DMF. Polymer solution with THF produces high density pores of fibers. The surface area increased. Polymer solution with DMF produced smooth fibers with almost complete loss of microtexture.

2.3.2.3 Solution Conductivity

Solution with higher conductivity has larger charge carrying capacity. The fiber jet of polymer solution which has higher conductivity will subject to greater tensile force. Zhang et al. (2005) has examined the effect of adding ions to PVA/water solution to increase the solution conductivity. They had added sodium chloride (NaCl) with increasing concentration to the PVA/water solution. In the result, the mean fiber diameter decreases. Hence, by increasing the concentration of NaCl ion increased the net charge density. It will then increase the electrical force of the jet. The fiber diameter will decrease (Table 2.1).

Table 2.1 Effects of electrospinning parameter on fiber morphology (Sill and von Recum 2008)

Parameter	Effect on fiber morphology
Applied voltage (increase)	Fiber diameter decreases initially then increases (not monotonic)
Flow rate (increase)	Fiber diameter increase (beaded morphologies occur if the flow rate is too high)
Distance between capillary tip and collector (increase)	Fiber diameter decreases (beaded morphologies occur if the distance between the capillary and collector is too short)
Polymer concentration (viscosity) (increase)	Fiber diameter increase (within optimal range)
Solution conductivity (increase)	Fiber diameter decreases (broad diameter distribution)
Solvent volatility (increase)	Fibers exhibit microtexture (pores on their surface, which increase surface area)

2.4 In Vitro Degradation of Scaffolds

Study of the hydrolytic degradation mechanism and rate is crucial factors for the application in biomedical and pharmaceutical applications. The degradation mechanism and rate of biodegradable polymeric scaffolds can be affected by numerous factors. The most common reasons for using absorbable polymer scaffolds are to accomplish time-varying mechanical properties and ensure complete dissolution of the implant, eliminating long-term biocompatibility concerns or avoiding secondary surgical operations. In order to release admixed materials such as antibiotics or growth factors, scaffold degradation may also be desired.

References

Annis, D., Bornat, A., Edwards, R., Higham, A., Loveday, B., & Wilson, J. (1978). An elastomeric vascular prosthesis. *ASAIO Journal, 24,* 209–214.

Anton, F. (1939). Method and apparatus for spinning. US Patent 2,160,962.

Baumgarten, P. K. (1971). Electrostatic spinning of acrylic microfibers. *Journal of Colloid and Interface Science, 36,* 71–79.

Dalton, P. D., Lleixà Calvet, J., Mourran, A., Klee, D., & Möller, M. (2006). Melt electrospinning of poly-(ethylene glycol-block-ε-caprolactone). *Biotechnology Journal, 1,* 998–1006.

Doshi, J., & Reneker, D. H. (1995). Electrospinning process and applications of electrospun fibers. *Journal of Electrostatics, 35,* 151–160.

Fisher, A., De Cossart, L., How, T., & Annis, D. (1984). Long term in-vivo performance of an electrostatically-spun small bore arterial prosthesis: The contribution of mechanical compliance and anti-platelet therapy. *Life Support Systems: The Journal of The European Society for Artificial Organs, 3,* 462–465.

Formhals, A. (1934). Process and apparatus for preparing artificial threads. US 1975504.

Formhals, A. (1940). Artificial thread and method of producing same. US 187306.

Gupta, P., & Wilkes, G. L. (2003). Some investigations on the fiber formation by utilizing a side-by-side bicomponent electrospinning approach. *Polymer, 44,* 6353–6359.

Huang, Z. M., He, C. L., Yang, A., Zhang, Y., Han, X. J., Yin, J., et al. (2006). Encapsulating drugs in biodegradable ultrafine fibers through co-axial electrospinning. *Journal of Biomedical Materials Research, Part A, 77*, 169–179.

Lam, C. X. F., Mo, X., Teoh, S.-H., & Hutmacher, D. (2002). Scaffold development using 3D printing with a starch-based polymer. *Materials Science and Engineering C, 20*, 49–56.

Lanza, R., Langer, R., & Vacanti, J. (2007). *Principles of tissue engineering*. London: Academic Press.

Larrondo, L., & St John Manley, R. (1981). Electrostatic fiber spinning from polymer melts. I. Experimental observations on fiber formation and properties. *Journal of Polymer Science: Polymer Physics Edition, 19*, 909–920.

Li, D., & Xia, Y. (2004). Electrospinning of nanofibers: Reinventing the wheel? *Advanced Materials, 16*, 1151–1170.

Liu, X., & Ma, P. X. (2004). Polymeric scaffolds for bone tissue engineering. *Annals of Biomedical Engineering, 32*, 477–486.

Loscertales, I. G., Barrero, A., Guerrero, I., Cortijo, R., Marquez, M., & Ganan-Calvo, A. (2002). Micro/nano encapsulation via electrified coaxial liquid jets. *Science, 295*, 1695–1698.

Ma, P. X. (2004). Scaffolds for tissue fabrication. *Materials Today, 7*, 30–40.

Megelski, S., Stephens, J. S., Chase, D. B., & Rabolt, J. F. (2002). Micro- and nanostructured surface morphology on electrospun polymer fibers. *Macromolecules, 35*, 8456–8466.

Reneker, D. H., & Chun, I. (1996). Nanometre diameter fibres of polymer, produced by electrospinning. *Nanotechnology, 7*, 216.

Sencadas, V., Ribeiro, C., Nunes-Pereira, J., Correia, V., & Lanceros-Méndez, S. (2012). Fiber average size and distribution dependence on the electrospinning parameters of poly(vinylidene fluoride–trifluoroethylene) membranes for biomedical applications. *Applied Physics A, 109*, 685–691.

Sill, T. J., & von Recum, H. A. (2008). Electrospinning: Applications in drug delivery and tissue engineering. *Biomaterials, 29*, 1989–2006.

Stitzel, J., Liu, J., Lee, S. J., Komura, M., Berry, J., Soker, S., et al. (2006). Controlled fabrication of a biological vascular substitute. *Biomaterials, 27*, 1088–1094.

Tan, E., Ng, S., & Lim, C. (2005). Tensile testing of a single ultrafine polymeric fiber. *Biomaterials, 26*, 1453–1456.

Taylor, G. (1969). Electrically driven jets. *Proceedings of the Royal Society of London, Series A: Mathematical and Physical Sciences, 313*, 453–475.

Whang, K., Thomas, C., Healy, K., & Nuber, G. (1995). A novel method to fabricate bioabsorbable scaffolds. *Polymer, 36*, 837–842.

Yu, D.-G., Zhu, L.-M., White, K., & Branford-White, C. (2009). Electrospun nanofiber-based drug delivery systems. *Health (1949-4998), 1*, 67–75.

Zhang, C., Yuan, X., Wu, L., Han, Y., & Sheng, J. (2005). Study on morphology of electrospun poly (vinyl alcohol) mats. *European Polymer Journal, 41*, 423–432.

Zhang, Y. Z., Wang, X., Feng, Y., Li, J., Lim, C. T., & Ramakrishna, S. (2006). Coaxial electrospinning of (fluorescein isothiocyanate-conjugated bovine serum albumin)-encapsulated poly(ε-caprolactone) nanofibers for sustained release. *Biomacromolecules, 7*, 1049–1057.

Zhao, F., Yin, Y., Lu, W. W., Leong, J. C., Zhang, W., Zhang, J., et al. (2002). Preparation and histological evaluation of biomimetic three-dimensional hydroxyapatite/chitosan-gelatin network composite scaffolds. *Biomaterials, 23*, 3227–3234.

Chapter 3
Fabrication of Polymer and Composite Scaffolds Using Electrospinning Techniques

Abstract This chapter reports the electrospinning technique for the formation of nano and microfibers. Due to the ability to fabricate fibrous scaffolds with micro and nano-scale properties, electrospinning technique has received much interest. Poly(caprolactone) (PCL) fibrous scaffolds with micro and nano-scale fibers and surface-porous fibers have not been explicitly investigated. In this study, the results of modulating the factors on processing route on nanofibrous scaffold morphology were investigated. 10 and 13 % w/v of PCL/dichloromethane (DCM) or chloroform was used at different flow rate and applied voltage. The result shows that 13 % w/v of PCL/chloroform produced better fibers. The fibrous scaffolds had two different ranges of fiber diameters. Average fiber diameter in the higher range was 4.52 μm while average fiber diameter in the lower range was 440 nm. In vitro degradation study suggested slow degradability of PCL electrospun fibers. This chapter also reports the fabrication of hydroxyapatite/PCL microfibers and their characteristics.

Keywords Electrospinning · Processing parameters · PCL polymers · Hydroxyapatite

3.1 Fabrication of Poly(Caprolactone) Nano-fibrous Scaffolds Using Electrospinning Technique

Over past few decades, there has been considerable interest in developing biodegradable nanofibrous scaffolds as effective drug delivery devices and tissue engineering applications. Many polymers have been used in drug delivery research to deliver the drug to targeted sites and increase the therapeutic benefits as well as minimize the side effects (Kreuter 1994; Sultana et al. 2014). Polymers that are used to fabricate scaffold have to be biodegradable and non toxic. There are a number of

Mohd Izzat Hassan, Mim Mim Lim and Naznin Sultana.

© The Author(s) 2015
N. Sultana et al., *Composite Synthetic Scaffolds for Tissue Engineering and Regenerative Medicine*, SpringerBriefs in Materials,
DOI 10.1007/978-3-319-09755-8_3

25

polymers that are used in the research of fabricating scaffold. These polymers include poly(lactic acid) (PLA), poly(caprolactone) (PCL), poly(glycolic acid) (PGA), poly(lactide-co-glycolide) (PLGA) and poly(hydroxybutyrate-co-hydroxy-valerate) (PHBV) (Sahoo et al. 2007; Bulasara et al. 2011). Thus, porous scaffolds that can prolong the release of bioactive factors are urgently required in tissue engineering (Sultana and Wang 2012).

Electrospinning technique is used widely for constructing fibrous tissue engineering scaffolds because electrospun fiber mimics the cellular microenvironment and enhances cell attachment as well as proliferation and differentiation (Doshi and Reneker 1995). Studies have shown that nanofibrous scaffolds have more advantages than non-nanofibrous scaffolds. Nanofibrous scaffolds enhance cell attachment, proliferation and differentiation (Bulasara et al. 2011). It is non-mechanical processing strategy that can be used to process a variety of native and synthetic polymers into highly porous materials composed of nano-scale to micron-scale diameter fibers (Roozbahani et al. 2013). Hence, Fig. 2.1 shows the schematic illustration of electrospinning technique. This technique uses an electric field to create a charged jet of polymer solution. Moreover, the jet is ejected and made to travel in the air and then collected on a grounded metal screen (Sill and Recum 2008). The product produced is called electrospun membrane or electrospun fiber. Electrospinning is a versatile process. It uses electrostatic force to produce fibers by electrospinning with polymer solution or polymer melt. The diameter of the polymer fibers produced ranges from a few nanometers to micrometers. When electrostatic force balances with surface tension, Taylor cone will be formed. Thus by increasing the electric field, fiber jet will be ejected from the apex of Taylor cone when electrostatic force is larger than surface tension. Therefore, during the ejection and elongation process, solvent is evaporated. The elongation and evaporation of the solvent reduces the diameter of the fiber from micrometer to nanometer. However, the electrified jet will undergo stretching and whipping process, while long and thin thread will be formed. Lastly, the polymer fiber produced is attracted by grounded collector (Sill and Recum 2008; Reneker and Chun 1996; Prabhakaran et al. 2008; Moghe et al. 2009).

PCL is synthetic aliphatic polyester which is hydrophobic (Roozbahani et al. 2013). Nowadays, PCL is often used in biomedical application because of its advantages ranging from low in cost, biocompatibility, biodegradability, and has good mechanical properties (Moghe et al. 2009). Moreover, PCL is an implantable device material approved by FDA. PCL is soluble in various solvents including chloroform, dichloromethane, carbon tetrachloride, benzene, toluene, cyclohexanone and 2-nitropropane at room temperature (Roozbahani et al. 2013).

In this study, solution parameters and processing parameters were optimized to fabricate PCL electrospun nano and microfibers. Characterization of the PCL electrospun fibers and in vitro degradation were investigated. The results suggested that the electrospun fibers could be used for tissue engineering applications.

PCL with molecular weight 70,000–90,000 were supplied from Sigma-Aldrich. Dichloromethane and chloroform were used as solvents. Firstly, 10 % w/v of PCL solution in DCM was prepared by dissolving 0.5 g of PCL in 5 ml of DCM.

Thereafter, the solution was stirred using magnetic stirrer at 600 rpm until all PCL dissolved. This was then repeated by changing the PCL concentration and solvent to chloroform. The polymer solution was electrospun using electrospinning unit (NaBond Nanofiber Electrospinning Unit, China). Parameters including polymer concentration, applied voltage and flow rate were adjusted and optimized to obtain a stable jet and uniform fiber. Nevertheless, capillary tip to collector distance was set constantly at 10 cm.

The morphology and structure of electrospun fibers was examined under a scanning electron microscope (SEM TM3000) and while fiber diameters were measured using Image J software. At least 30 fibers diameters were measured from SEM images and thus the average fiber diameter was calculated. The fiber diameter was expressed as mean ± SD, while their distributions were exhibited as histograms. Drug incorporated PCL electrospun membrane was immersed into phosphate buffer saline (PBS) in a water bath at 37 °C for 4 months. After 23, 53 and 107 days, the PCL electrospun membrane was taken out, dried in room temperature and observed under SEM.

Obtained data were expressed as mean ± SD. Using Student's t-test, statistical analyses were carried out and a p-value <0.05 was considered to be statistically significant. Electrospun fibers were fabricated as in the parameters in Table 3.1.

Solvent plays an important role in producing better fiber morphology. In this study, electrospinning was unsuccessful when DCM was used as solvent. The reason for this is the low boiling point of DCM which is only 40 °C. However, when high voltage was applied, the solvent evaporates and leaves polymer. Furthermore, there is no ejection of the as a result of low connectivity of the solvent. Nevertheless, electrostatic force is unable to overcome the surface tension. Hence, a good solvent has to be chosen in order to produce better fibers. Figure 3.1 shows the results of using 10 % w/v polymer concentration with chloroform as solvent, flow rate of 1.5, 2 and 3 ml/h were applied. During electrospinning process, these jets were not stable. There were no fiber structures in the SEM images. It is due to low polymer concentration. Polymer concentration influences both viscosity and surface tension of the solution (Reneker and Chun 1996). If the viscosity of the polymer is

Table 3.1 The parameters to fabricate PCL electrospun fibrous scaffold

PCL (% w/v)	Solvent	Flow rate (ml/h)	Applied voltage (kV)	Observation
10	DCM	0.1	25.00	Unsuccessful
15	DCM	0.1	26.26	Unsuccessful
10	Chloroform	1.5	15.16	Unsuccessful
10	Chloroform	2.0	11.68	Unsuccessful
10	Chloroform	3.0	20.03	Unsuccessful
13	Chloroform	1.0	24.00	Unsuccessful
13	Chloroform	0.8	10.13	Unsuccessful
13	Chloroform	0.5	10.06	Successful
13	Chloroform	0.3	8.32	Successful
13	Chloroform	0.1	7.47	Successful

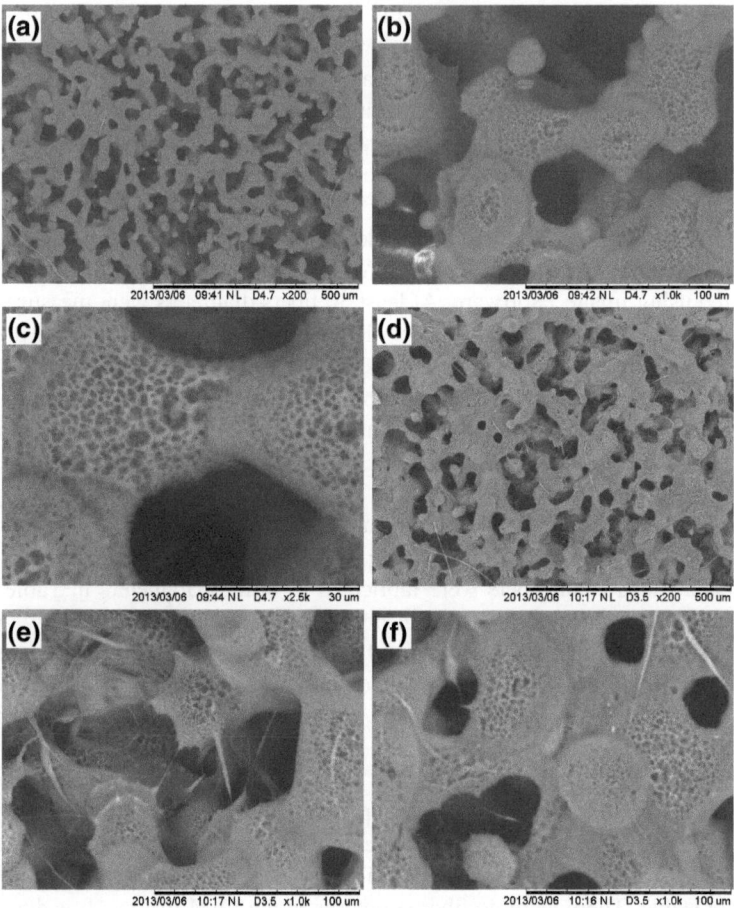

Fig. 3.1 SEM images of 10 % w/v PCL in chloroform: **a** 1.5 ml/h flow rate with magnification 200; **b** 1.5 ml/h flow rate with magnification 1.0 k; **c** 1.5 ml/h flow rate with magnification 2.5 k; **d** 2.0 ml/h flow rate with magnification 200; **e** 2.0 ml/h flow rate with magnification 1.0 k; **f** 2.0 ml/ h flow rate with magnification 1.0 k

too low, entanglements between polymer chains become unable to provide a stable jet. Hence, polymer fibers will break up into droplets before reaching the collector due to the effect of surface tension. However, pores were formed and shown in the SEM images of higher magnification.

The influence of increasing polymer concentration was studied on the morphology of electrospun fibers. PCL polymer concentration was increased to 13 % w/v in chloroform. Different flow rates were applied including 0.1, 0.3, 0.5, 0.8 and 1.0 ml/h. Flow rate of 0.8 and 1.0 ml/h were too high for 13 % w/v PCL solution as unstable fiber jets were formed and the fibers produced were not uniform. Flow rate 0.1, 0.3 and 0.5 ml/h formed better fibers as shown in Fig. 3.2. Fewer surface-pores were formed in fibers of flow rate 0.5 ml/h and the fiber jet

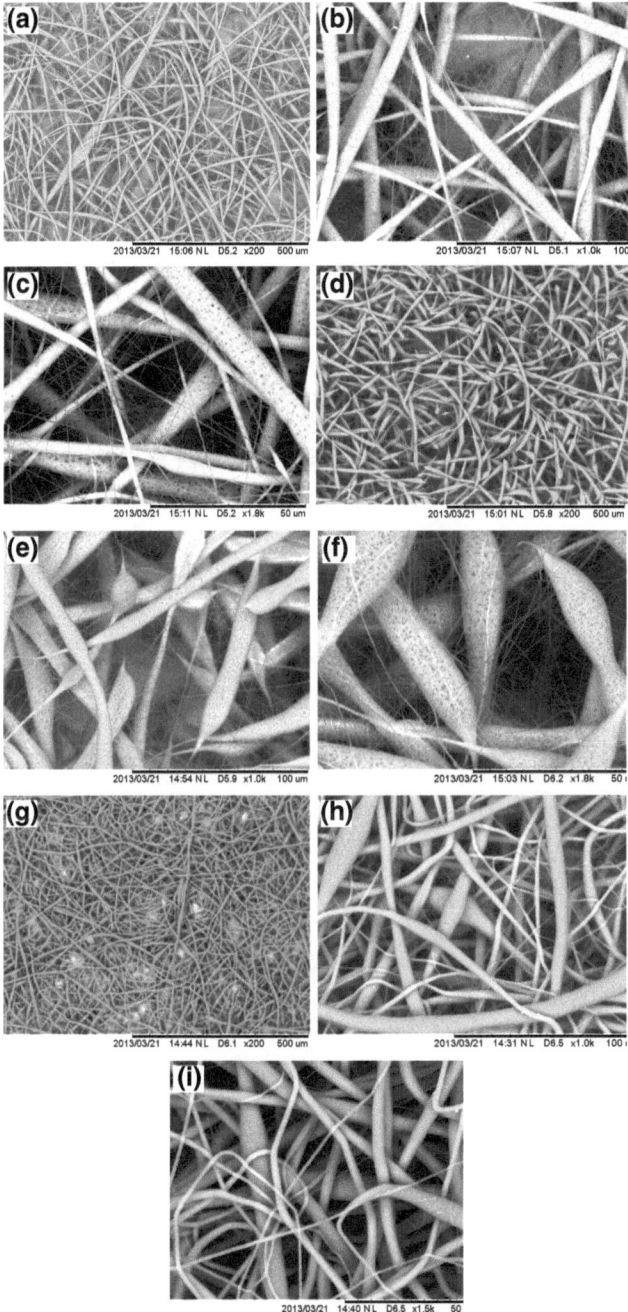

Fig. 3.2 SEM image of 13 %/v PCL in chloroform: **a** 0.1 ml/h flow rate with magnification 200; **b** 0.1 ml/h flow rate with magnification 1.0 k; **c** 0.1 ml/h flow rate with magnification 1.8 k; **d** 0.3 ml/h flow rate with magnification 200; **e** 0.3 ml/h flow rate with magnification 1.0 k; **f** 0.3 ml/h flow rate with magnification 1.8 k; **g** 0.5 ml/h flow rate with magnification 200; **h** 0.5 ml/h flow rate with magnification 1.0 k; **i** 0.5 ml/h flow rate with magnification 1.5 k

splits. For flow rate 0.3 ml/h, there was some splitting on fiber jet and more surface-pores were formed. For flow rate of 0.1 ml/h, fiber jet formed was stable and uniform when applied voltage was set at 7.47 kV as well as capillary tip to collector distance at 10 cm. Among these parameters, parameters of flow rate 0.1 and 0.3 ml/h produced better fiber morphology with surface-pores. Thus, the reason for forming surface-pores was due to the quick evaporation of solvent. Nevertheless, surface-pores are important as it allows nutrients and waste flow as well as to regenerate ECM in vivo.

Polymer flow rate had an influence on fiber diameter. However, it was also observed that decreasing flow rate increased the pore size as pores were more visible at sample of 0.3 and 0.1 ml/h flow rate. Average diameter of sample 8 with flow rate 0.5 ml/h was 5.49 μm in higher range and 390 nm in lower range (Fig. 3.3a, b). Average diameter of sample 9 with flow rate 0.3 ml/h was 8.93 μm in higher range and 430 nm in lower range (Fig. 3.3c, d). Average diameter of sample 10 with flow rate 0.1 ml/h was 4.52 μm in higher range and 440 nm in lower range

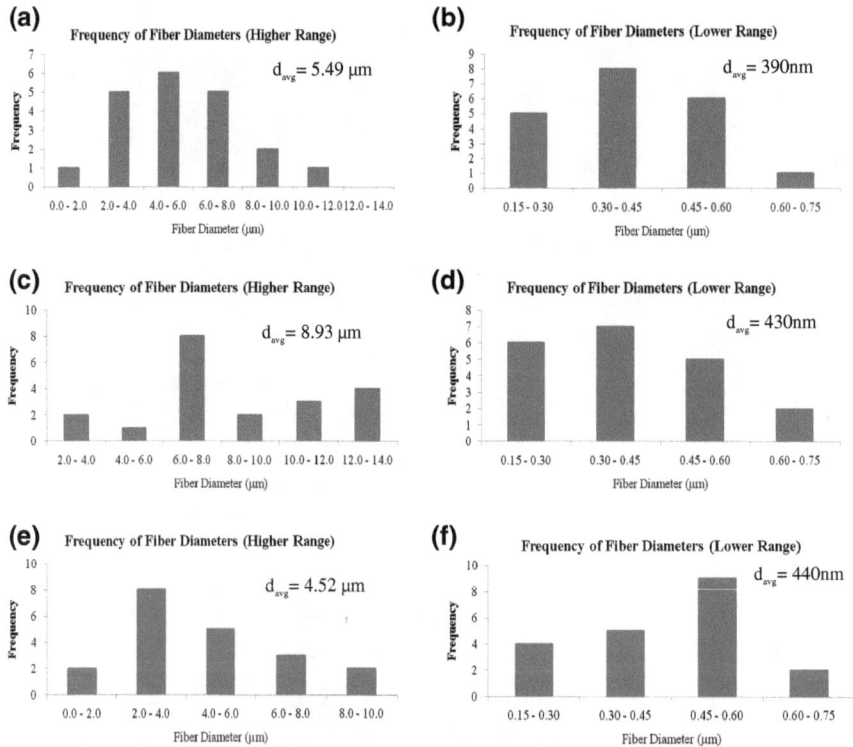

Fig. 3.3 Distribution of electrospun fiber diameters of 13 %/v PCL in chloroform with flow rate of 0.5 ml/h: **a** higher range; **b** lower range. Distribution of electrospun fiber diameters of 13 %/v PCL in chloroform with flow rate of 0.3 ml/h: **c** higher range; **d** lower range. Distribution of electrospun fiber diameters of 13 %/v PCL in chloroform with flow rate of 0.1 ml/h: **e** higher range; **f** lower range

(Fig. 3.3e, f). Moreover, the fiber diameter of higher range decreased but fiber diameter of lower range increased while decreasing the flow rate. It is believed that the applied voltage influences the fiber diameter.

Based on this experiment, the processing parameters of electrospun PCL fibers that gave suitable morphology include 13 %/v PCL solution in chloroform, 0.1/0.3 ml/h flow rate, 7.47/8.32 kV applied voltage, capillary tip to collector distance of 10 cm. These parameters formed a more stable jet and produced better fiber morphology.

As-fabricated and degraded electrospun fibers were compared using SEM (Fig. 3.4). It was observed that the samples did not contain broken fibers but the fibers became more random after degradation. Figure 3.5 shows the frequency of fiber diameters of sample after degraded for 30, 53 and 107 days. The average fiber diameter was 340 nm after degraded for 30 days, 330 nm after degraded for 53 days and 240 nm after degraded for 107 days. On the other hand, the average fiber diameter of as fabricated PCL fibers was 350 nm. Electrospun fibers are slowly degraded and PCL has slow degradation rate (Moghe et al. 2009). Due to significant slower degradation rate, PCL is more suitable to be designed as long term

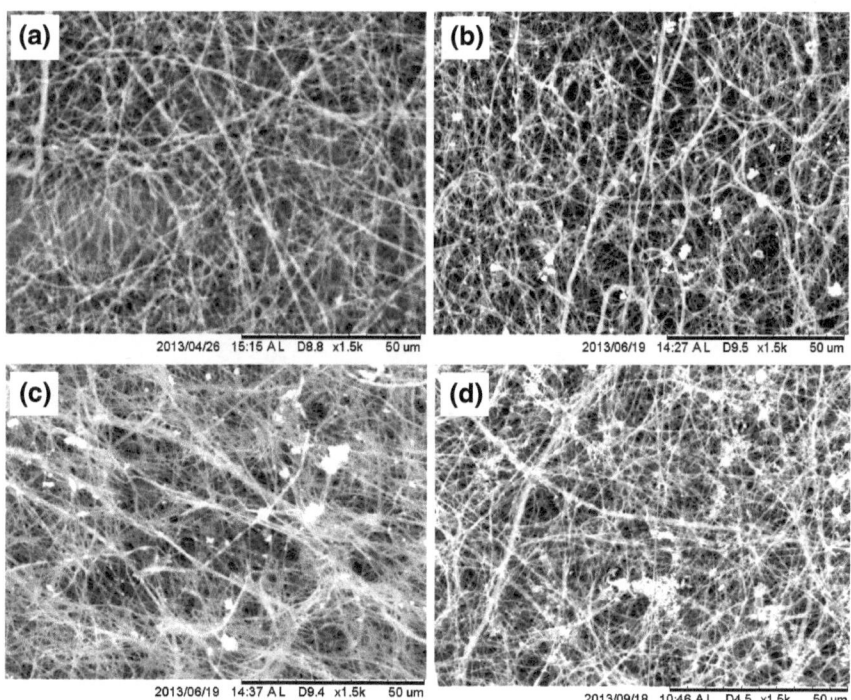

Fig. 3.4 SEM image of electrospun fibers of sample 13 %/v PCL in chloroform with flow rate 0.3 ml/h in magnification 1,500. **a** As fabricated; **b** after degraded for 30 days; **c** after degraded for 53 days; **d** after degraded for 107 days

implantable systems such as 1 year implantable contraceptive device or tissue engineering scaffolds.

Although production of electrospun fibers has gained prominence due to its versatility, there are some parameters that can influence the structure and morphology of the electrospun fibers produced which include processing parameters and solution parameters. Processing parameters are applied voltage, flow rate and capillary tip to collector distance. However, increase in the applied voltage will lead to the formation of Taylor cone. Further increase in the voltage will result in the ejection of fiber jet from the apex of the Taylor cone. Megelski et al. investigated polystyrene/tetrahydrofuran (THF) solution on flow rate parameter. The result showed that increasing flow rate will increase fiber diameter and pore size. However, at high flow rate, bead defects will form as a result of insufficient evaporation of solvent. The fiber is not completely dry before reaching collector. The increase of flow rate increases the volume from the needle tip. This will lead to an increase in evaporation time of the solvent and larger polymer crystallization time. Hence, a

Fig. 3.5 Distribution of electrospun fiber diameters of sample 13 %/v PCL in chloroform after degraded for **a** 30 days; **b** 53 days; **c** 107 days

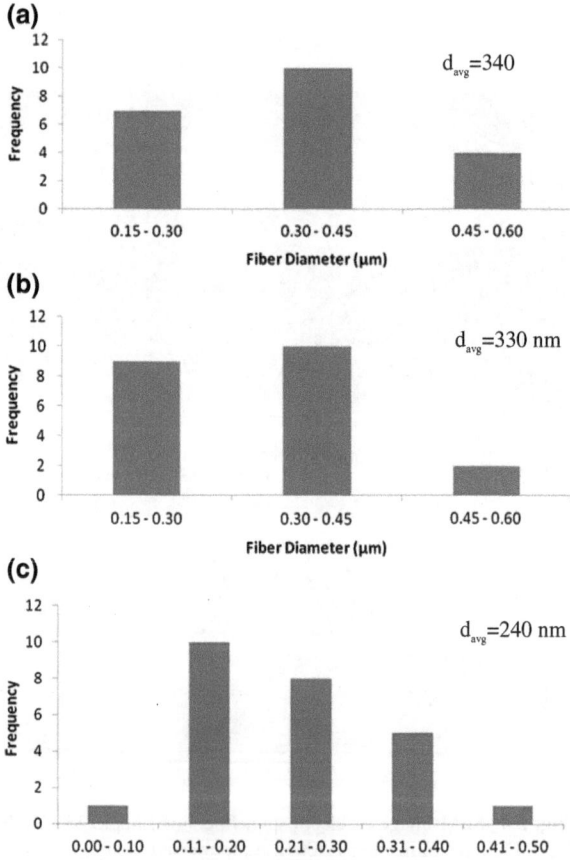

larger diameter of fiber and broader size distribution will be produced (Moghe et al. 2009). During fiber jet ejection, evaporation of the solvent occurs. Fiber need to be completely dried before reaching the collector to avoid formation of bead defects. Hence, the distance between capillary tip and grounded collector plays an important role to achieve a good fiber structure. Fiber diameter decreases with increasing distance between capillary tip and collector. In addition, the formation of beaded electrospun polystyrene fibers due to shortening of the travelling distance which is attributed to the inadequate drying of polymer fiber prior to reaching the collector. Thus, solution parameters include solvent volatility, solution viscosity and polymer concentration. Volatile solvent is used due to phase separation during the ejection of jet. Polymer concentration will affect the spin ability of the solution. If the concentration is too dilute, fibers will break up into droplets. Therefore, within an optimum range of polymer concentration, fiber diameter increases with the increase of polymer concentration. Fibers cannot be formed in too concentrated polymer solution. The solution will block the needle outlet. With higher solution viscosity, larger fiber diameter will be formed (Reneker and Chun 1996).

It was reported that PCL and its mixture with polylactide (PLA) were electro-spun from chloroform-acetone mixture produced electrospun fibers of average diameters of 2 μm (Valle et al. 2011). It was also reported that 10 % w/v PCL in methylene chloride/dimethyl formamide (MC/DMF) produced nanofibrous web where some micro sized fibers were embedded in nano sized fibers (Kanani and Bahrami 2011). Electro-wet spinning technique was reported to be able to produce porous electrospun fibers of average fiber diameters 0.5–12 μm (Khil et al. 2005). In present study, we found the ability of using electrospinning technique to produce two different ranges of fiber diameters which were in micrometer and nanometer sizes using another solvent. This phenomenon is called bimodal distribution of fiber size. Nowadays, researchers are trying to obtain bimodal morphology in biological scaffolds (Gholipour et al. 2009; Huang et al. 2003; Beachley and Wen 2009; Pham et al. 2006; Baji et al. 2010) as it resemble natural extracellular matrix and has high biological effects in biomedical applications (Kanani and Bahrami 2011). 10 wt% polyethylene oxide in aqueous solution was able to form a bimodal distribution of fiber size (Deitzel et al. 2001). Moreover, bimodal nanofibers were formed when suitable amount of graphene oxide was blended with nylon-6 solution (Pant et al. 2012). In our study, we found the ability of 13 % w/v PCL with chloroform as solvent at 0.1–0.5 ml/h flow rate to form porous and bimodal distribution of average fiber diameter in the range of 4–9 μm and around 400 nm (Fig. 3.5). We have reported the degradation characteristics as well.

3.1.1 Conclusions

In this study, using the electrospinning technique, PCL fibrous scaffolds with micro and nanoscale fiber diameter were formed. A number of parameters had significant effects on fiber diameter and morphology. This study shows that dependence of different parameters to produce different morphology of electrospun fibers depending on various applications. No significant degradation was observed during the drug release period.

3.2 Fabrication of Electrospun Hydroxyapatite/ Polycaprolactone Scaffolds

Tissue engineering fibrous scaffolds serves as three-dimensional (3D) environmental framework by mimicking the extracellular matrix (ECM) for cells to grow. Biodegradable polycaprolactone (PCL) microfibers were fabricated to mimic the ECM as a scaffold with 7.5 and 12.5 % w/v concentrations. At lower PCL concentration of 7.5 % w/v resulted microfibers with bead defects. The average diameter of fibers increased at higher voltage and the distance of tip to collector. Further investigation of the incorporation of nano sized hydroxyapatite (nHA) into nanofiber was performed. The incorporation of 10 % w/w nHA with 7.5 % w/v PCL solution produced submicron sized beadless fibers. Biodegradable PCL and nHA/PCL could be promising for tissue engineering scaffold application.

The recent advanced approach for replacing lost tissue, damaged organ and end-stage of organ failure has been known as tissue engineering. Scaffold or matrix is one of the elements of tissue engineering. It serves as framework for cells by providing appropriate environment to live and grow, promoting an extracellular matrix and other biological molecules synthesization, and by facilitating the formation of tissue and organ function. Various techniques have been investigated, such as freeze-drying, phase-separation, solvent casting or particulate leaching, gas foaming, electrospinning, and rapid prototyping (Sultana and Khan 2013a, b; Mou et al. 2013; Ji et al. 2012; Schueren et al. 2011; Serra et al. 2013).

In the recent years, electrospinning has gained wide interest among the researchers for producing fibers with controlled diameter, high surface area and porous structure. It has been utilized for producing polymeric nanofibers. The technique is able to produce continuous nanofiber by applying an electrostatic field to a polymer solution driven by high voltage supply between a needle tip and the collector. The advantages of electrospinning are the production of uniform and broad range of diameter of nanofibers with long and continuous characteristic compared to other techniques (Schueren et al. 2011).

Biodegradable and biocompatible synthetic polymer polycaprolactone (PCL) has slower degradation rate and is a suitable candidate for fabricating fibrous scaffolds. It has been approved by food and drug administration (FDA) for various medical applications. PCL can be dissolved in many solvents such as chloroform, dimethylformamide (DMF), hexafluoroisopropanol (HFIP), dichloromethane (DCM), and tetrafluoroethylene (TFE) (Serra et al. 2013; Shalumon et al. 2011; Zoppe et al. 2009; Simşek et al. 2012; Nguyen et al. 2013; Chen et al. 2013). However, PCL is hydrophobic. On the other hand, bioceramics such as hydroxyapatite (HA) and calcium phosphate can be incorporated with PCL to produce composite electrospun fibers through electrospinning. HA is osteoconductive and hydrophilic than PCL polymers. Incorporation of nHA into PCL fibers could have better properties than PCL fibers alone.

The objectives of present work were to use the electrospinning technique to produce beadless or defect free PCL and nHA/PCL composite fibers. The effects of polymer solution concentration and other parameters of electrospinning technique on the morphology of PCL microfibers and nHA/PCL composite microfibers were also investigated.

Polycaprolactone (M_w: 70,000–90,000), was purchased from Sigma-Aldrich. 99.8 % acetone was purchased from the same company. Hydroxyapatite nanoparticle was produced in-house from mixing acetone solution of $Ca(NO_3)_2 \cdot 4H_2O$ with an aqueous solution of $(NH_4)_2HPO_4$ and NH_4HCO_3 by nanoemulsion technique reported in the previous study (Hassan et al. 2012; Zhou et al. 2008).

The PCL pellets was weighted and dissolved in acetone under magnetic stirrer at 40 °C. nHA/PCL solution was prepared by adding 10 % w/w of nHA into 7.5 % w/v PCL solution. The solution was then homogenized using a homogenizer IKA T25 (IKA Works) in order to disperse HA nanoparticles in the PCL polymer solution.

The experiment of electrospinning was carried out by using an electrospinning unit (NaBond, China). The PCL and nHA/PCL solutions were transferred into a 5 ml syringe with a blunt-end needle with 18G and 22G in size. The feeding rate of the injection syringe was fixed at 1, 1.5, 2 and 3 ml/h using an infusion pump (Veryark TCV-IV) while the distance between the collector plate and the tip of blunt-end needle was set to 10 and 15 cm. The tip of blunt needle was connected to the positive electrode of high-voltage power supply while negative electrode was connected to a metallic rack as the grounded target collector. The metallic rack was wrapped with an aluminium foil (10×10 cm^2) as a collector for collecting electrospun fibers. Electrospun fibers was ejected from the syringe onto the collector by varying high voltage DC power supply in the range of 0–30 kV. The collected electrospun membrane was then stored in a dessicator prior to analysis.

In the beginning of the experiment of the electrospinning process, the tip of the blunt-needle was observed to check any droplet of the solution. When the droplet of the solution became into a large glob shape, the droplet at the tip of the needle was quickly wiped off using tissue and cotton bud to prevent clogging from the needle tip. The voltage was slowly increased and stopped until a stable Taylor cone was achieved. As the potential increased, the droplet emerged and move towards the direction of the collector caused by electrostatic force between the needle-tip and the collector.

The electrospun fibers was examined under a scanning electron microscopy (SEM; Hitachi TM-3000, Japan) and a field emission scanning electron microscopy (FESEM). The fibers were cut into a square section, mounted on aluminium stub and observed at an accelerating voltage of 15 kV. The diameters of resulting fibers were measured at random location on each fiber. At least 50 measurements from the images were recorded using image analysis software (Image J, NIH, USA). The samples were sputter coated with gold before analysis under FESEM. Energy dispersive X-ray (EDX) attached with the SEM was used to confirm the presence of HA in the PCL polymer solution.

Using PCL solution in acetone, PCL fibers were obtained by electrospinning. Figure 3.6 shows the SEM micrographs of electrospun PCL microfibers at different PCL concentrations. Microfibers with 7.5 % w/v concentration contained beads in the fibers while the microfibers with 12.5 % w/v concentration had homogeneous and continuous fibers. The morphology of the fiber produced was adversely affected by the lower concentration of the PCL.

Although the PCL fibers of 12.5 % w/v concentration had smooth and continuous microstructure, however the diameter of the fiber was wider than that of the microfibers produced from 7.5 % w/v polymer concentration. The diameter of microfibers produced from 12.5 % w/v was in the range of 1–10 μm. On the other hand, fibers produced from 7.5 % w/v PCL concentration had the diameter ranging from 100 nm to 5 μm. This was in agreement with another study where it was described that lower concentration (5 and 7.5 % w/v) produced nano-size fibers with beading and higher concentration (10 % w/v) produced micron-size fiber without beading (Bosworth and Downes 2012). The problem with lower concentration was the resulting nanofibers had spindle-like bead characteristic. The concentration gives some trade-off between the fiber size and existence of beading along with the fibers. The beading defect as in Fig. 3.6a–c is the common problem which needs to be overcome for producing fibers by electrospinning. Particularly large beads with most of spindle-like structure were obtained with an average size of 3.85 and 3.60 μm for each 15 and 22 kV. The beads size was in range of micrometer scale with porous surface. When the distance between the grounded collector and the needle tip increased, the average fiber diameter also increased. The average fiber diameter increased with the decreased flow rate for 12.5 % w/v concentration.

Figure 3.7 shows the dependence of fiber diameter on different polymer concentration. The flow rate and the distance were crucial in determining the size of fiber diameter. When using lower flow rate, the applied voltage was increased to achieve stable Taylor cone. Thus the fiber diameter with larger size was resulted. After increasing the distance between the needle tip and the collector by 5 cm at low concentration, the fiber diameter was slightly increased.

The results based on the parameters given in Table 3.2, it was observed that there are many parameters are needed to be adjusted to obtain nanofibers. Chloroform is commonly used solvent for electrospinning of PCL but the diameter of fiber produced was in micrometer. However, some of researchers successfully obtained fibers in nanometer scale by combining with other solvent such as chloroform/dimethylformamide, chloroform/methanol, and chloroform/acetone (Meng et al. 2010; Amna et al. 2013). Acetone is less harmful compared to other solvents, thus was explored in this study. However, in this study beadless or defect free fibers resulted in micro-scale sizes.

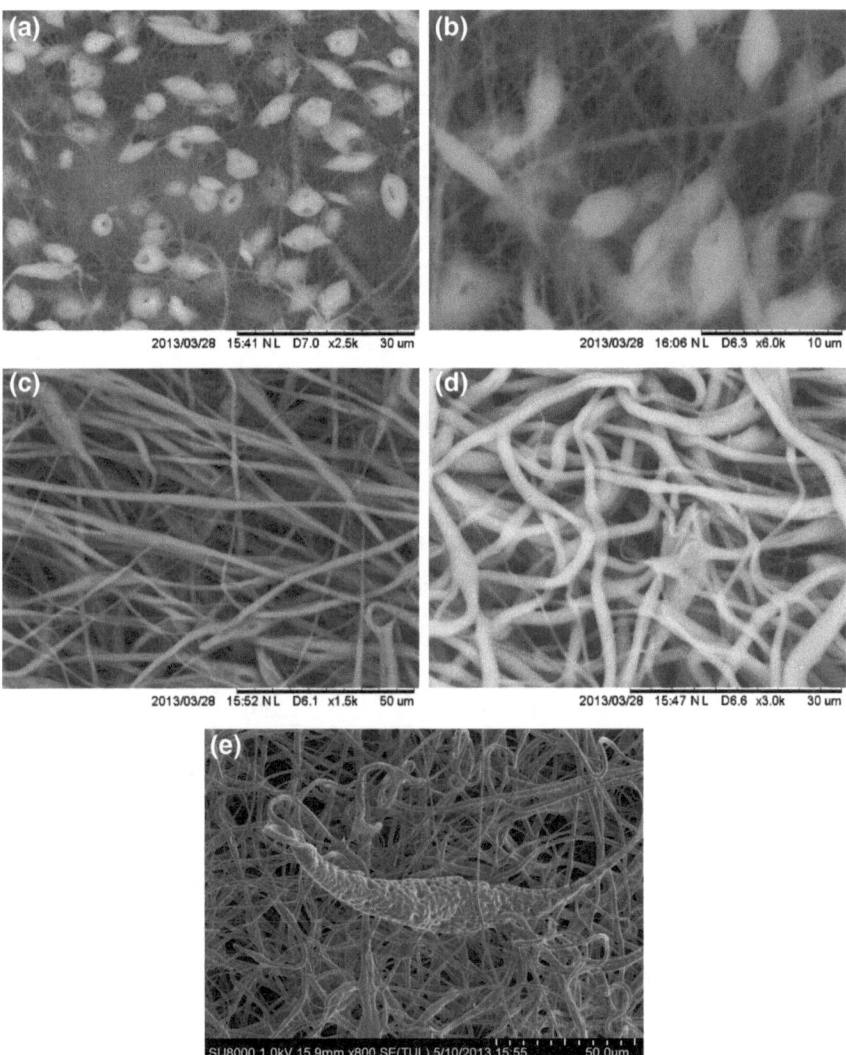

Fig. 3.6 SEM micrographs of PCL fibers with 7.5 % w/v at **a** 10 cm and **b** 15 cm tip-collector distance while 12.5 % w/v at **c** 15 kV, 3 ml/h and **d** 22 kV, 1.5 ml/h; **e** FESEM micrograph of bead defect at 7.5 % w/v PCL (Hassan et al. 2014)

Hydroxyapatite nanoparticles (nHA) were incorporated into PCL solution of 7.5 % w/v concentration. At constant diameter and voltage, six different samples were electrospun at 1, 2, and 3 flow rates (ml/h) using 18G and 22G blunt needle. We observed that nHA was successfully incorporated along the fiber as shown in Fig. 3.8. SEM micrographs confirmed that nHA was well-mixed with PCL polymer solution and nHA was embedded well into PCL fibers. The nHA were

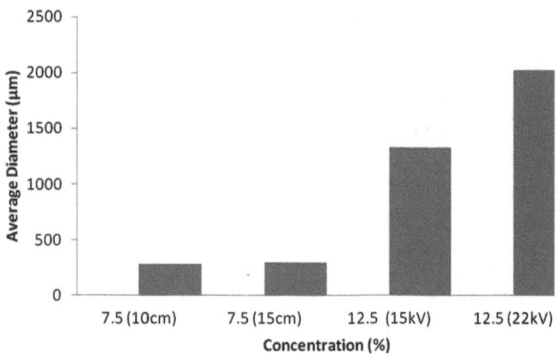

Fig. 3.7 Average diameters (μm) of fibers at different concentration (%) of PCL with vary distances from tip of needle to the collector (cm), and applied voltage (kV)

Table 3.2 Electrospinning parameters for PCL fiber (Hassan et al. 2014)

Concentration (% w/v)	Diameter (cm)	Voltage (kV)	Flow rate (ml/h)	Average diameter (nm)	Characteristic
7.5	10	15	3	287.27	Beaded
7.5	15	15	3	302.91	Beaded
12.5	10	15	3	1,330	Smooth
12.5	10	22	1.5	2,030	Smooth

homogenously distributed within the fibers possibly due to the low concentration of nHA. If lower concentration of nHA is used, it is possible to overcome the agglomeration of nanoparticles. Using a homogenizer can be used to disperse the nHA particles throughout the fibers. The average diameter size of fibers was not affected by the different parameters of flow rate and gauge size. The average diameter of fibers was in one micrometer size. Interestingly, the incorporation of nHA made the nHA/PCL fibers smoother with almost unnoticeable bead defects than the electrospun PCL fibers at the same concentration. However, an increase in fiber diameter was observed (from nanometer to submicron size). The result was in contrast with another study where the fiber diameter was observed to be decreased after adding nHA into PCL (Croisier et al. 2012; Jaiswal et al. 2013). However, they used another solvent to produce electrospun HA/PCL nanofibers and it is well known in literature that the solvent plays an important role in fiber diameter sizes (Table 3.3).

The elemental mapping further confirmed the distribution of nHA in PCL nanofiber matrices (Fig. 3.9). EDX analyses at different locations of composite electrospun fibers also validated the presence of nHA particles that were well mixed with polymer fiber surfaces (Fig. 3.9). The success of electrospinning of polymeric

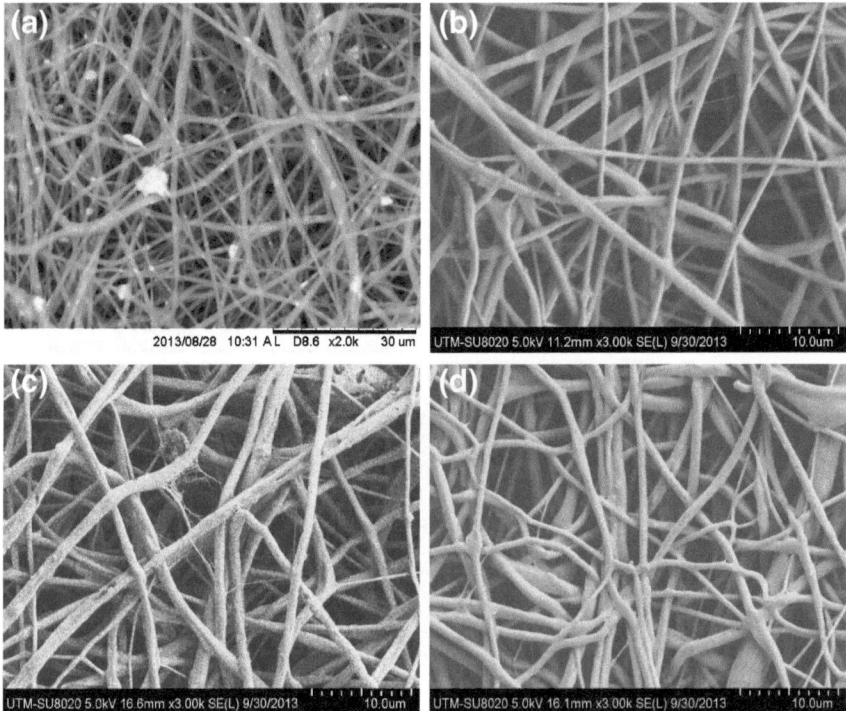

Fig. 3.8 SEM micrograph of 10 % HA in 7.5 % w/v of nHA/PCL fibers at: **a** 1 ml/h and 22 kV; FESEM micrographs at 22 kV and flow rates of **b** 3 ml/h, **c** 2 ml/h, and **d** 1 ml/h, respectively (Hassan et al. 2014)

Table 3.3 Electrospinning parameters for PCL/HA fiber (Hassan et al. 2014)

Concentration (% w/v)	Voltage (kV)	Flow rate (ml/h)	Gauge size (G)	Average diameter (μm)
7.5	22	3	18	1.25
7.5	22	3	22	1.11
7.5	22	2	18	1.18
7.5	22	2	22	1.23
7.5	22	1	18	1.44
7.5	22	1	22	1.16

fibrous scaffolds lies in the process of preparing the suitable solution concentration and the process of electrospinning itself. The applied voltage, flow rate and distance between the tip and the collector does effect the distribution of diameter of obtained fiber. However, producing fibers without bead defects is also needed.

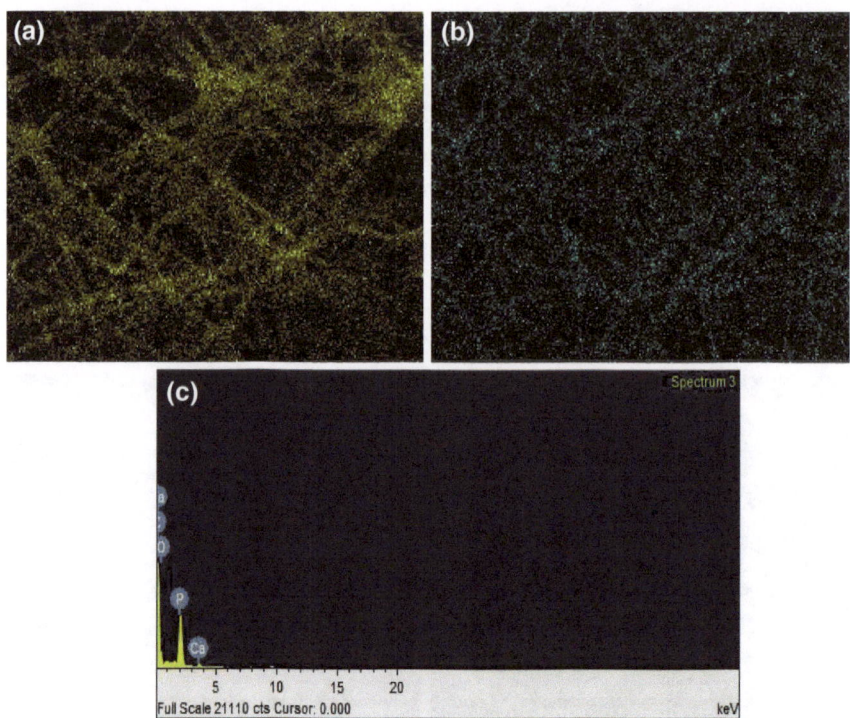

Fig. 3.9 Elemental mapping of **a** calcium and **b** phosphorus, and **c** EDX spectrum of 7.5 % w/v PCL at 1 ml/h, 22 kV. (Hassan et al. 2014)

3.2.1 Conclusions

PCL nanofibers were obtained at lower concentration of 7.5 % w/v PCL polymer concentration. At higher concentration of 12.5 % w/v, fibers of submicron scale was produced. Incorporation of nHA produced beadless fibers without agglomeration. However, the incorporation of nHA made the fibers thicker than that of PCL nanofibers. Further study on bioactivity of the fabricated electrospun fibers, in vitro biological assessment are still in progress. The electrospun nHA/PCL is expected to be conducive for cell growth as it contains osteoconductive nHA that has the bone bonding ability to be used as tissue engineering fibrous scaffolds.

Acknowledgment The authors would like to acknowledge the Faculty of Biosciences and Medical Engineering, Universiti Teknologi Malaysia (UTM) for the lab facilities. This work was supported by research grants FRGS (vot no: 4F126), GUP Tier 1 (03 H13, 05H07). Authors also acknowledge the support provided by MOHE, RMC and UTM.

References

Amna, T., Barakat, N. A. M., Hassan, M. S., Khil, M.-S., & Kim, H. Y. (2013). Camptothecin loaded poly(ε-caprolactone)nanofibers via one-step electrospinning and their cytotoxicity impact. *Colloids and Surfaces A: Physicochemical and Engineering Aspects, 431*, 1–8.

Baji, A., Mai, Y.-W., Wong, S.-C., Abtahi, M., & Chen, P. (2010). Electrospinning of polymer nanofibers: Effects on oriented morphology, structures and tensile properties. *Composites Science and Technology, 70*, 703–718.

Beachley, V., & Wen, X. (2009). Effect of electrospinning parameters on the nanofiber diameter and length. *Materials Science and Engineering: C, 29*, 663–668.

Bosworth, L. A., & Downes, S. (2012). Acetone, a sustainable solvent for electrospinning poly(ε-caprolactone) fibres: Effect of varying parameters and solution concentrations on fibre diameter. *Journal of Polymers and the Environment, 20*, 879–886.

Bulasara, I. K., Uppaluri, R., & Purkait, M. K. (2011). Manufacture of nickel-ceramic composite membranes in agitated electroless plating baths. *Materials and Manufacturing Processes, 26*, 862–867.

Chen, Z., Cao, L., Wang, L., Zhu, H., & Jiang, H. (2013). Effect of fiber structure on the properties of the electrospun hybrid membranes composed of poly(ε-caprolactone) and gelatin. *Journal of Applied Polymer Science, 127*, 4225–4232.

Croisier, F., Duwez, A. S., Jérôme, C., Léonard, A. F., Vanderwerf, K. O., Dijkstra, P. J., et al. (2012). Mechanical testing of electrospun PCL fibers. *Acta Biomaterialia, 8*, 218–224.

Deitzel, J. M., Jkleinmeyer, J., Harris, D., & Tan, N. C. B. (2001). The effect of processing variables on the morphology of electrospun nanofibers and textiles. *Polymer, 42*, 261–272.

Doshi, J., & Reneker, D. H. (1995). Electrospinning process and application of electrospun fibers. *Journal of Electrostatics, 35*, 151–160.

Gholipour, A., Bahrami, S. H., & Nouri, M. (2009). Chitosan-poly (vinyl alcohol) blend nanofibers: Morphology, biological and antimicrobial properties. *e-Polymers, 9*, 1580–1591.

Hassan, M. I., Mokhtar, M., Sultana, N., & Khan, T. H. (2012). Production of hydroxyapatite (HA) nanoparticle and HA/PCL tissue engineering scaffolds for bone tissue engineering. In *IEEE* (pp. 239–242).

Hassan, M. I., Sun, T., & Sultana, N. (2014). Fabrication of nano hydroxyapatite/poly (caprolactone) composite microfibers using electrospinning technique for tissue engineering applications. *Journal of Nanomaterials, 2014*, 1–7.

Huang, Z.-M., Zhang, Y.-Z., Kotaki, M., & Ramakrishna, S. (2003). A review on polymer nanofibers by electrospinning and their applications in nanocomposites. *Composites Science and Technology, 63*, 2223–2253.

Jaiswal, A. K., Chhabra, H., Kadam, S. S., Londhe, K., Soni, V. P., & Bellare, J. R. (2013). Hardystonite improves biocompatibility and strength of electrospun polycaprolactone nanofibers over hydroxyapatite: A comparative study. *Materials Science & Engineering C, Materials for Biological Applications, 33*, 2926–2936.

Ji, C., Annabi, N., Hosseinkhani, M., Sivaloganathan, S., & Dehghani, F. (2012). Fabrication of poly-DL-lactide/polyethylene glycol scaffolds using the gas foaming technique. *Acta Biomaterialia, 8*, 570–578.

Kanani, A. G., & Bahrami, S. H. (2011). Effect of changing solvents on poly(ε-caprolactone) nanofibrous webs morphology. *Journal of Nanomaterials, 2011*, 31.

Khil, M.-S., Bhattarai, S. R., Kim, H.-Y., Kim, S.-Z., & Lee, K.-H. (2005). Novel fabricated matrix via electrospinning for tissue engineering. *Journal of Biomedical Materials Research Part B: Applied Biomaterials, 72B*, 117–124.

Kreuter, J. (1994). Drug targeting with nanoparticles. *European Journal of Drug Metabolism andPharmacokinetics, 19*(3), 253–256.

Meng, Z. X., Zheng, W., Li, L., & Zheng, Y. F. (2010). Fabrication and characterization of three-dimensional nanofiber membrance of PCL–MWCNTs by electrospinning. *Materials Science and Engineering: C, 30*, 1014–1021.

Moghe, A. K., Hufenus, R., Hudson, S. M., & Gupta, B. S. (2009). Effect of the addition of a fugitive salt on electrospinnability of poly(ε-caprolactone). *Polymer, 50*, 3311–3318.

Mou, Z.-L., Duan, L.-M., & Zhang, Z.-Q. (2013). Preparation of silk fibroin/collagen/ hydroxyapatite composite scaffold by particulate leaching method. *Materials Letters, 105*, 189–191.

Nguyen, T.-H., Bao, T. Q., Park, I., & Lee, B.-T. (2013). A novel fibrous scaffold composed of electrospun porous poly (epsilon-caprolactone) fibers for bone tissue engineering. *Journal of Biomaterials Applications, 28*, 514–528.

Pant, H. R., Park, C. H., Tijing, L. D., Amarjargai, A., Lee, D.-H., & Kim, S. K. (2012). Bimodal fiber diameter distributed graphene oxide/nylon-6 composite nanofibrous mats via electrospinning. *Colloids and Surfaces A: Physicochemical and Engineering Aspects, 407*, 121–125.

Pham, Q. P., Sharma, U., & Mikos, A. G. (2006). Electrospun poly(ε-caprolactone) microfiber and multilayer nanofiber/microfiber scaffolds: Characterization of scaffolds and measurement of cellular infiltration. *Biomacromolecules, 7*, 2796–2805.

Prabhakaran, M. P., Venugopal, J. R., Chyan, T. T., Hai, L. B., Chan, C. K., Lim, A. Y., et al. (2008). Electrospun biocomposite nanofibrous scaffolds for neural tissue engineering. *Tissue Engineering Part A, 14*, 1787–1797.

Reneker, D. H., & Chun, I. (1996). Nanometre diameter fibres of polymer, produced by electrospinning. *Nanotechnology, 7*, 216–223.

Roozbahani, F., Sultana, N., Ismail, A. F., & Nouparvar, H. (2013). Effects of chitosan alkali pretreatment on the preparation of electrospun PCL/chitosan blend nanofibrous scaffolds for tissue engineering application. *Journal of Nanomaterials, 2013*, 6.

Sahoo, N. G., Jung, Y. C., & Cho, J. W. (2007). Electroactive shape memory effect of polyurethane composites filled with carbon nanotubes and conducting polymer. *Materials and Manufacturing Processes, 22*, 419–423.

Schueren, L. V. D., Schoenmaker, B. D., Kalaoglu, O. I., & Clerck, K. D. (2011). An alternative solvent system for steady state electro spinning of polycaprolactone. *European Polymer Journal, 47*, 1256–1263.

Serra, T., Planell, J. A., & Navarro, M. (2013). High-resolution PLA-based composite scaffolds via 3-D printing technology. *Acta Biomaterialia, 9*, 5521–5530.

Shalumon, K. T., Anulekha, K. H., Chennazhi, K. P., Tamura, H., Nair, S. V., & Jayakumar, R. (2011). Fabrication of chitosan/poly (caprolactone) nanofibrous scaffold for bone and skin tissue engineering. *International Journal of Biological Macromolecules, 48*, 571–576.

Sill, T. J., & Recum, H. A. V. (2008). Electrospinning: Applications in drug delivery and tissue engineering. *Biomaterials, 29*, 1989–2006.

Simşek, M., Capkın, M., Karakeçili, A., & Gümüşderelioğlu, M. (2012). Chitosan and polycaprolactone membranes patterned via electrospinning: Effect of underlying chemistry and pattern characteristics on epithelial/fibroblastic cell behavior. *Journal of Biomedical Materials Research Part A, 100A*, 3332–3343.

Sultana, N., & Khan, T. H. (2013a). Polycaprolactone scaffolds and hydroxyapatite/polycaprolactone composite scaffolds for bone tissue engineering. *Journal of Bionanoscience, 7*, 169–173.

Sultana, N., & Khan, T. H. (2013b). Water absorption and diffusion characteristics of nanohydroxyapatite (nHA) and poly(hydroxybutyrate-co-hydroxyvalerate-) based composite tissue engineering scaffolds and nonporous thin films. *Journal of Nanomaterials, 2013*, 1.

Sultana, N., & Wang, M. (2012). PHBV/PLLA-based composite scaffolds fabricated using an emulsion freezing/freeze-drying technique for bone tissue engineering: Surface modification and in vitro biological evaluation. *Biofabrication, 4*, 015003.

Sultana, N., Mokhtar, M., Hassan, M. I., Jin, R. M., Roozbahani, F., & Khan, T. H. (2014). Chitosan-based nanocomposite scaffolds for tissue engineering applications. *Materials and Manufacturing Processes*.

Valle, L. J., Camps, R., Díaz, A., Franco, L., Rodríguez-Galán, A., & Puiggalí, J. (2011). Electrospinning of polylactide and polycaprolactone mixtures for preparation of materials with tunable drug release properties. *Journal of Polymer Research, 18*, 1903–1917.

Zhou, W. Y., Wang, M., Cheung, W. L., Guo, B. C., & Jia, D. C. (2008). Synthesis of carbonated hydroxyapatite nanospheres through nanoemulsion. *Journal of Materials Science: Materials in Medicine, 19,* 103–110.

Zoppe, J. O., Peresin, M. S., Habibi, Y., Venditti, R. A., & Rojas, O. J. (2009). Reinforcing poly (epsilon-caprolactone) nanofibers with cellulose nanocrystals. *ACS Applied Materials & Interfaces, 1,* 1996–2004.

Chapter 4
Fabrication and Characterization of Polymer and Composite Scaffolds Using Freeze-Drying Technique

Abstract This chapter reports the emulsion freezing/freeze-drying technique for the formation of three dimensional scaffolds. Composite scaffolds based on biodegradable natural polymer and osteoconductive hydroxyapatite (HA) nanoparticles can be promising for a variety of tissue engineering (TE) applications. This study addressed the fabrication of three dimensional (3D) porous composite scaffolds composed of HA and chitosan fabricated via thermally induced phase separation and freeze-drying technique. The scaffolds produced were subsequently characterized in terms of microstructure, porosity, mechanical property. In vitro degradation and in vitro biological evaluation were also investigated. The scaffolds were highly porous and had interconnected pore structures. The pore sizes ranged from several microns to a few hundred microns. The incorporated HA nanoparticles were well mixed and physically co-existed with chitosan in composite scaffold structures. The addition of 10 % (w/w) HA nanoparticles into chitosan enhanced the compressive mechanical properties of composite scaffold compared to pure chitosan scaffold. In vitro degradation results in phosphate buffered saline (PBS) showed slower uptake properties of composite scaffolds. Moreover, the scaffolds showed positive response to mouse fibroblast L929 cells attachment. Overall, the findings suggest that HA/chitosan composite scaffolds could be suitable for TE applications.

Keywords Composite · Scaffolds · Chitosan · Hydroxyapatite

4.1 Hydroxyapatite/Chitosan Scaffolds Using Freeze-Drying Technique

The development of three-dimensional (3-D) scaffold which can provide an appropriate microenvironment for tissue growth and regeneration is required by tissue engineering (TE) (Lanza et al. 2007; Griffith 2002). TE aims at finding

Mohd Izzat Hassan and Naznin Sultana.

© The Author(s) 2015
N. Sultana et al., *Composite Synthetic Scaffolds for Tissue Engineering and Regenerative Medicine*, SpringerBriefs in Materials,
DOI 10.1007/978-3-319-09755-8_4

biodegradable and biocompatible scaffolds that can be seeded with cells for both in vitro and in vivo purposes (Atala and Lanza 2002). One of the major objectives of the scaffolds is to mimic the natural characteristics of extracellular matrix (ECM). The appropriate scaffold architecture is vital to allow vascularization and to supply adequate nutrients to the developing tissue. Therefore, scaffolds must possess some special features, such as, an appropriate surface chemistry, biodegradability, non-toxicity, an interconnected porous network, appropriate pore size and shape to attain sufficient nutrient transport and cell in-growth (Hollister 2005; Khan et al. 2008). Thermally induced phase separation (TIPS) of polymer solution was reported to be used in the field of drug delivery and to fabricate microspheres in order to incorporate biological and pharmaceutical agents (Ma 2004; Chen and Ma 2005). In order to fabricate tissue engineering scaffolds, this process has become much popular. By altering the types of polymer and solvent, polymer concentration and phase separation temperature, different types of porous scaffolds with micro and macro-structured scaffolds can be produced. In order to control pore morphology on a micrometer to nanometer level, TIPS process can be utilized (Ma 2004). In this technique, firstly, the polymer is dissolved in a solvent, and then the bioactive molecule is dissolved or dispersed in the resulting homogeneous solution, which is then cooled in a controlled fashion until solid-liquid or liquid-liquid phase separation is induced. The resulting bicontinuous polymer and solvent phases are then quenched to create a two phase solid. By sublimation, solidified solvent is then removed, leaving a porous polymer scaffold with bioactive molecules incorporated within the polymer. To incorporate small molecules into the polymer scaffolds, this technique is useful.

Chitosan is one of the most common natural biopolymers used for bone tissue engineering (Han et al. 2010; Wu et al. 2011; Vandevord et al. 2002). Natural polymers can possess highly organized structure to guide cells to grow and may stimulate an immune response at the same time. Being a linear polysaccharide, chitosan is composed of glucosamine and N-acetyl glucosamine units. Due to several properties such as tissue compatibility, bioresorbability, antibacterial activity and haemostatic characteristics, chitosan is a suitable material for bio-medical applications (Rinaudo 2006). Moreover, the degradation products of chitosan are non-toxic, non-immunogenic, non-carcinogenic (Wu et al 2012). On the other hand, hydroxyapatite (HA) is a frequent choice and possesses osteocon-ductive properties which demonstrated excellent cellular and tissue responses in vitro and in vivo. Together with biocompatibility, HA have chemical structures similar to bone minerals and have been used as bone graft, augmentation and substitution (Park and Bronzino 2003; Ratner 2004; Sultana and Wang 2008a; Tong et al. 2010).

In this study, chitosan and HA/Chitosan scaffolds were fabricated with a desired pore sizes and porosity using thermally induced phase separation technique. In a second step, the scaffolds were characterized using various methods. The in vitro degradation and the response of fibroblast cells on porous chitosan-based scaffolds were also evaluated.

Chitosan with medium molecular weight and in the powder form were purchased from Sigma-Aldrich. 2 % (v/v) acetic acid solution was prepared using glacial acetic acid in ultra-pure milliQ water. The HA nanoparticles were produced via a nanoemulsion process (Zhou et al. 2008). The thermally induced phase separation technique was used to fabricate the scaffold. Firstly, 0.5 g of chitosan was weighed accurately into a centrifuge tube. Then an accurately measured amount of 2 % acetic acid was added to the centrifuge tube to make an emulsion. The mixtures were homogenized using a homogenizer (Ultra-Turrax, IKAWERKE, Germany) at a fixed speed. The homogeneous solution was put into a capped glass tube and frozen in a freezer at a preset temperature of −18 °C for 24 h. The frozen samples were then freeze-dried for 72 h in freeze-drying vessel (LABCONCO-Freeze dry system, USA) to remove the solvent completely. The scaffolds produced were stored in a desiccator until it being used for characterization.

The HA/chitosan composite scaffolds were also fabricated using the thermally induced phase separation technique. The HA nanoparticles were weighed accurately and added to the centrifuge tube containing 0.5 g of chitosan polymer. Then 20 ml of 2 % (v/v) acetic acid was added, homogenized using the homogenizer (Ultra-Turrax, IKAWERKE, Germany) and transferred into a freezer at a preset temperature in capped glass tubes. The solidified solution was maintained overnight and then transferred into a freeze-drying vessel (LABCONCO-Freeze dry system, USA) for several days to remove the solvent. Thus, HA/Chitosan porous scaffolds were obtained and stored in a desiccator until it being used for characterization.

The cylindrical porous scaffolds were cut into specimens with the size of 15 mm in diameter and 1.5 mm in height using a sharp razor blade. The morphology of the scaffolds was studied using a scanning electron microscope (SEM: Hitachi TM 3000). The presence and distribution of HA nanoparticles in the HA/Chitosan composite scaffolds were analyzed by energy dispersive X-ray (EDX) spectrometry. Using the liquid displacement method, the density and porosity of the scaffolds were measured (Hsu et al. 1997; Sultana and Wang 2008b). Using an Instron mechanical tester, the compressive mechanical properties of the scaffolds were determined at a crosshead speed of 0.5 mm/min with 100 N load cell.

The specimens of diameter (10 mm) and height (1.5 mm) were cut from chitosan and 10 % (w/w) HA incorporated HA/chitosan composite scaffolds fabricated from 2.5 % polymer solution and weighed. Phosphate buffered saline (PBS) was prepared by dissolving PBS (supplied by Sigma) with distilled water. The specimens were placed in sealable vials containing 10 ml of PBS solution (pH 7.4) and PBS solution was replaced with new solution each week. The samples from PBS were removed at regular intervals, rinsed with distilled water and dried using a freeze dryer and weighed. The experiment was performed for 4 weeks.

Weight loss during investigation was determined as:

$$Weight\ loss\ (\%) = (W_i - W_f)/W_i \times 100 \tag{2.1}$$

where W_i and W_f are specimen weights before and after soaking in PBS.

The water uptake was calculated using the following equation:

$$Water\ uptake\ (\%) = (W_w - W_d)/W_d \times 100 \qquad (2.2)$$

where W_d and W_w are specimen weights before and after soaking in PBS.

The morphologies of the as fabricated and degraded composite scaffolds and thin films were studied with a scanning electron microscopy (SEM: Hitachi TM 3000) at 12 kV.

All data were presented as mean ± standard deviations (SD). To test the significance, an unpaired student's t test (two-tail) was applied and a value of $p < 0.05$ was considered to be statistically significant.

Figure 4.1a shows freeze-dried HA nanoparticles produced in-house using a nano-emulsion process. Figure 4.1b represents scaffolds samples produced through thermally induced phase separation process. The scaffolds were relatively large in size and homogeneous in appearance. They could be handled easily and normally did not contain voids (viz., macropores with sizes greater than 1 mm). The scaffolds fabricated had a nonporous thin skin layer.

Figure 4.2 is the SEM micrographs, revealing the porous structure of the interior of a scaffold sample. The processing parameters were found to have significant influence on the quality of chitosan scaffolds. From the microstructure examination by SEM, it was observed that the concentration of polymer solution had a large effect on the morphology of the scaffold. The polymer concentration on scaffold pore sizes and pore walls was evaluated. Scaffolds produced from solutions at the 2 % (w/v) chitosan concentration had thin pore walls and exhibited weak structures (Fig. 4.2a). Concentration higher than 3.3 % (w/v) possessed high viscosity, which ultimately prevented adequate homogenization. Scaffolds produced from 3.3 % (w/v) chitosan solution exhibited thick pore walls with smaller pore sizes (Fig. 4.2c). Scaffolds produced from 2.5 % (w/v) chitosan concentration had better porous structures (Figs. 4.2b and 4.3). At this concentration, porous structures were obtained with pore sizes over few hundred microns and reasonable thickness of pore walls (Fig. 4.3).

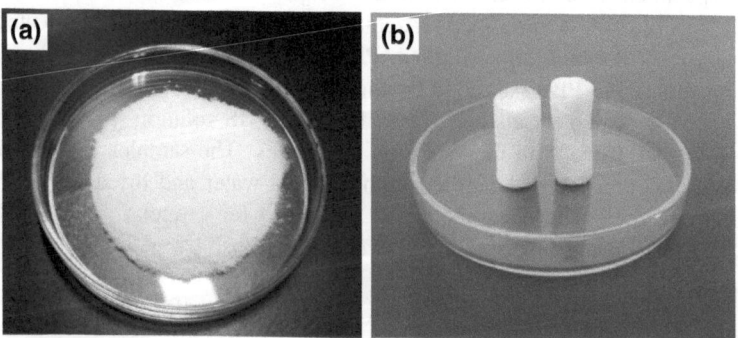

Fig. 4.1 General appearance of **a** HA nanoparticles; **b** chitosan and HA/chitosan scaffolds (Sun et al. 2014)

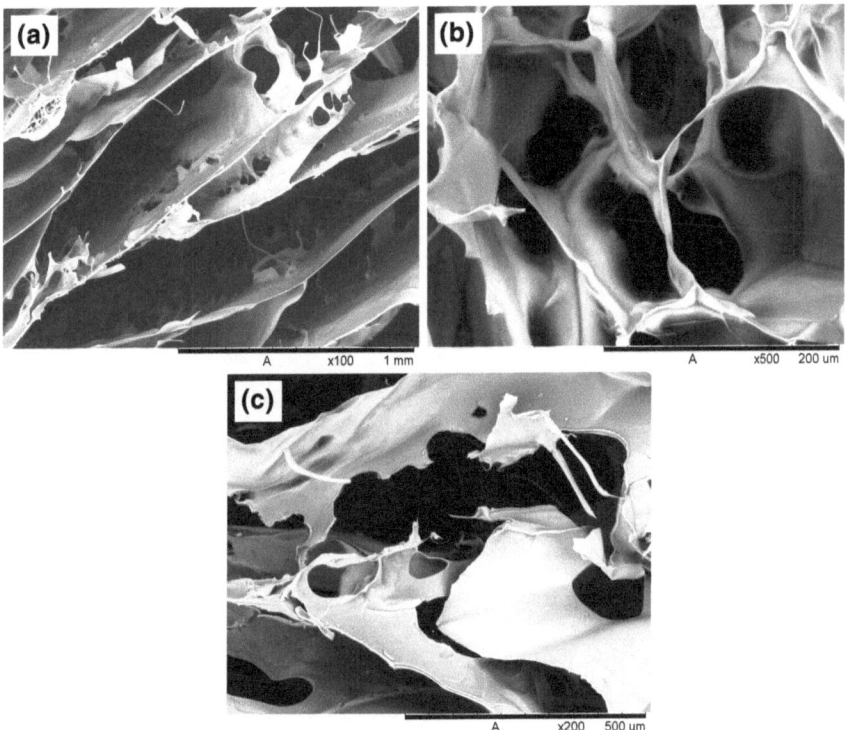

Fig. 4.2 SEM micrographs of chitosan scaffolds fabricated with different chitosan concentrations: **a** 2 %, **b** 2.5 %, **c** 3.3 % (ref. denote **a**, **b**, **c**) (Sun et al. 2014)

Figure 4.4 shows SEM micrographs of HA/chitosan composite scaffold. With the incorporation of 10 % of HA shown in Fig. 4.4a, the porous structure of the composite scaffolds did not change significantly. Compared to the pure chitosan scaffolds, the pore sizes of composite scaffolds decreased slightly (Fig. 4.5). It was observed that good distribution and good adhesion of HA nanoparticles in the chitosan matrix were present. Tiny agglomeration of HA was observed in the polymer pore walls. It was also observed that homogenizing at high speed reduced the agglomeration and helped the HA nanoparticles to disperse well in the polymer pore walls. Figure 4.6 is the EDX analyses at the different locations of composite scaffolds which confirmed the presence of HA particles inside the pore walls.

Chitosan and HA/chitosan scaffolds with high porosity were fabricated using 2.5 % (w/v) chitosan concentrations. With the incorporation of HA scaffolds density increased from 0.1783 to 0.2918 g/cm^3 and the porosity decreased from 88 to 82 % (Table 4.1). The pores were remained open for both chitosan and HA/chitosan scaffolds. The pore sizes were ranging from 20 to 350 μm. In another investigation that used a different polymer reported that by careful selection of different processing parameters, the pore sizes and the thickness of pore walls can

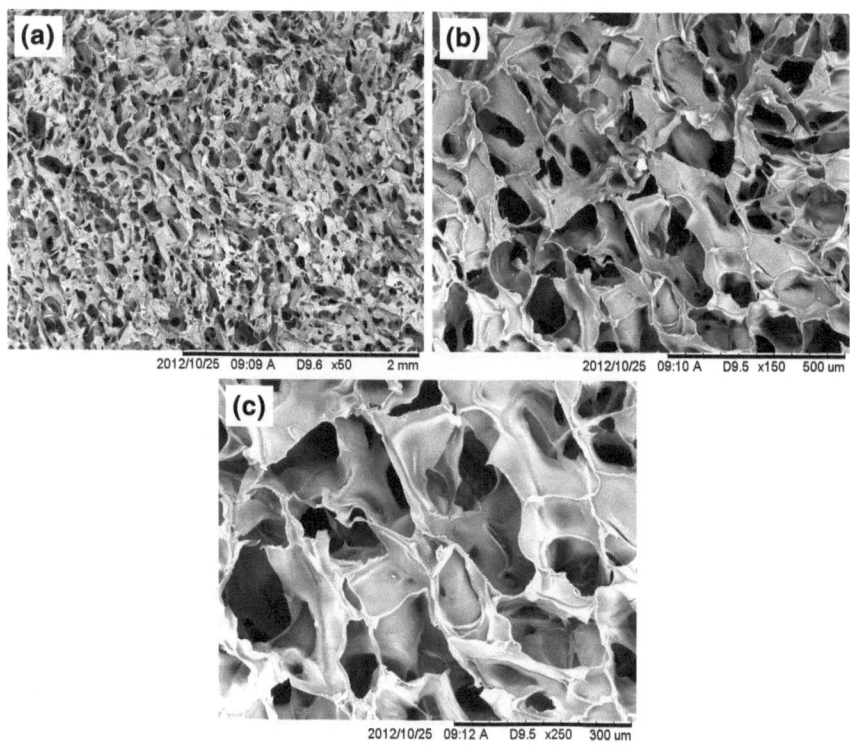

Fig. 4.3 SEM micrographs of chitosan scaffold at different magnification fabricated from 2.5 % chitosan concentration: **a** X50, **b** X150, **c** X250 (Sun et al. 2014)

Fig. 4.4 SEM micrographs of HA/chitosan composite scaffolds: **a** X150, **b** X250 (Sun et al. 2014)

be controlled (Sultana and Wang 2008a, 2012; Schugens et al. 1996). The formation of porous structure depends on the crystallization of the solvent phase when the solution temperature was lowered. During the phase separation at lower

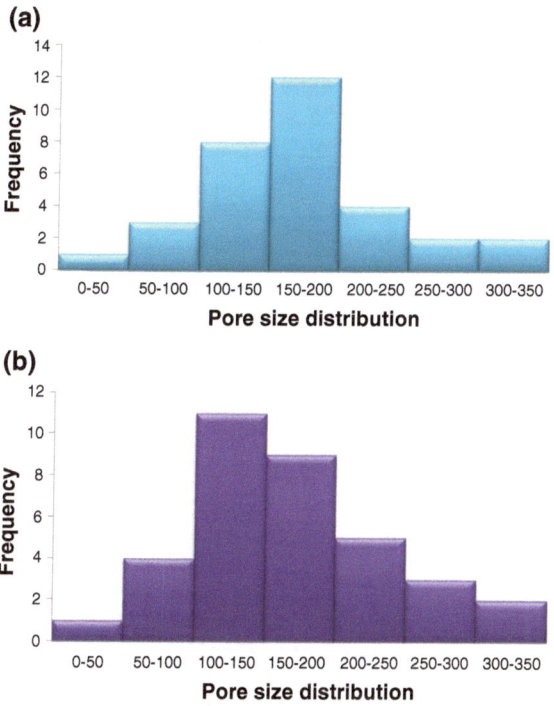

Fig. 4.5 Pore size distribution of **a** chitosan and **b** HA/chitosan scaffolds (Sun et al. 2014)

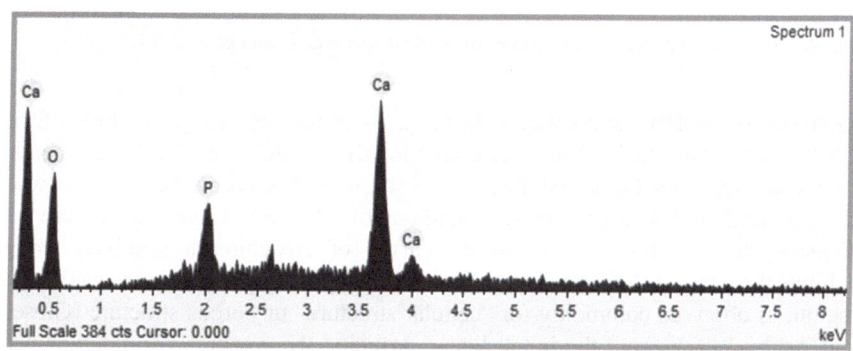

Fig. 4.6 An EDX spectrum of a HA/chitosan composite scaffold (Sun et al. 2014)

temperature, polymer phase was excluded from the solvent crystallization front and a continuous polymer-rich phase was formed. After the sublimation of solvent phase, scaffolds were formed with the pores of the same geometry of the solvent crystals. Similar phenomenon was also observed when the composite scaffolds were fabricated. Compressive properties of chitosan scaffolds increased with

Table 4.1 The density, porosity, pore type and compressive modulus of the scaffolds (Sun et al. 2014)

Scaffolds (w/v)	Density (g/cm³)	Porosity (%)	Pore type	Compressive modulus (MPa)
Chitosan	0.1783	88	Open	1.1
10 % HA/chitosan	0.2918	82	Open	2.8

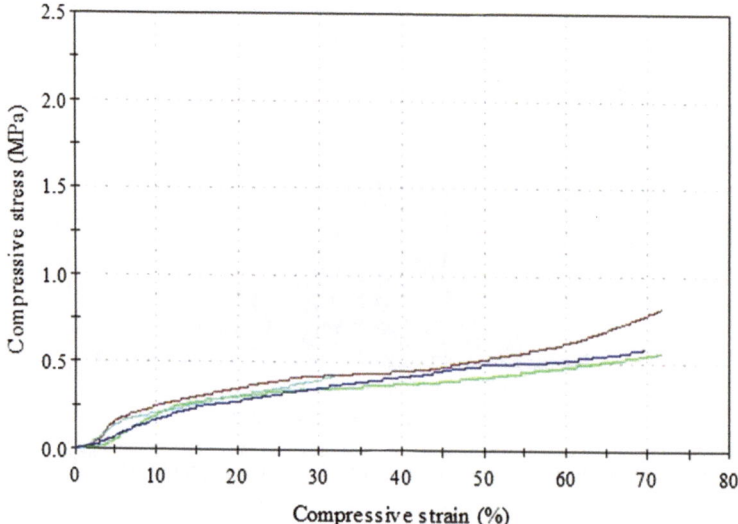

Fig. 4.7 Compressive stress-strain curves of scaffold specimens (Sun et al. 2014)

incorporation of HA nanoparticles. In the 2.5–7.5 % strain range, scaffolds from 2.5 % (w/v) chitosan scaffold had a compressive modulus of 1.1 MPa whereas composite scaffolds fabricated from 10 % (w/w) HA incorporated 2.5 % (w/v) chitosan scaffold had a compressive modulus of 2.8 MPa (Table 4.1). Figure 4.7 displays the compressive stress-strain curves of HA/chitosan scaffolds which exhibits three regions namely, initial linear elasticity, long plateau and densification region, as observed commonly for "cellular structure" or porous structure (Gibson and Ashby 1997). From the initial linear elasticity, the compressive modulus was calculated. As the HA/chitosan composite scaffolds had higher relative density, it had higher compressive modulus than that of pure chitosan scaffolds.

Figure 4.8a shows the comparison of the water uptake curves between chitosan scaffold and 10 % HA in chitosan composite scaffolds at 37 °C. It was observed that the initial water uptake of polymer scaffold was much higher than that of composite scaffold as chitosan is well known as hydrophilic polymer. The water uptake of pure chitosan scaffold was about 400 % on the other hand the water uptake of HA/PHBV composite scaffold was about 125 %. The addition of HA nanoparticles

Fig. 4.8 a Uptake of (*1*) HA/chitosan and (*2*) chitosan scaffolds; **b** weight loss of (*1*) chitosan and (*2*) HA/chitosan scaffolds after in vitro degradation for 4 weeks (Sun et al. 2014)

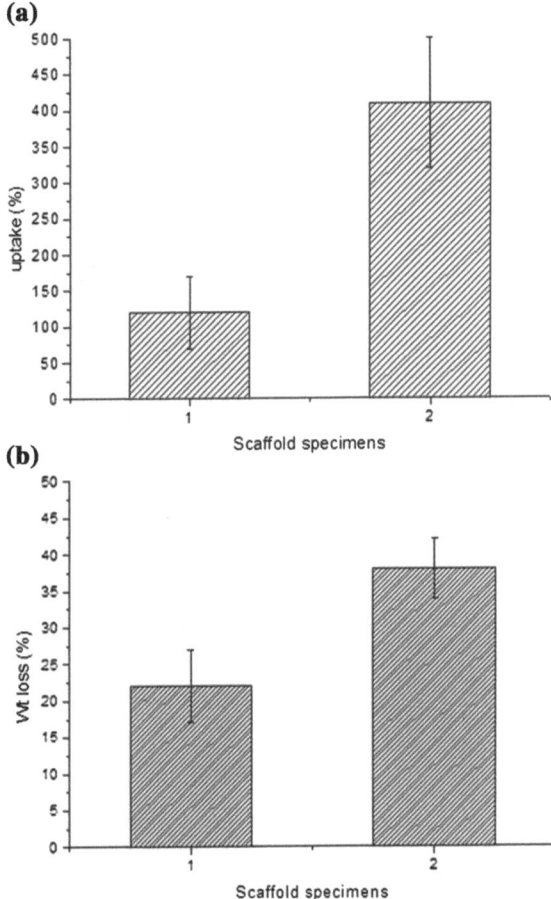

reduced the water uptake properties in the composite scaffolds by stiffening the matrix. The uptake properties were not only related to the degree of porosity as the micro-structure of porous matrix may also have an important role. It was reported that the pore size and pore geometry also influenced water uptake (van Vlierberghe et al. 2006). Figure 4.8b displays the weight loss of the scaffold specimens of chitosan (2.5 % w/v) and the same composition containing 10 % HA over 4 week period of time. After 4 week, the 10 % HA containing chitosan composite scaffold showed elevated weight loss (approx. 38 %) than polymer scaffolds (21 %). Accelerated weight loss of HA/chitosan composite scaffolds was mainly due to dissolution of HA in the scaffolds matrix.

Figure 4.9a, b shows the morphology of chitosan and HA/chitosan scaffolds after in vitro degradation in PBS at 37 °C for 4 weeks. Large pores were observed in the chitosan scaffolds after 4 weeks (Fig. 4.9a). Figure 4.9b shows the morphological changes of HA/chitosan composite scaffolds containing 10 % of HA after

Fig. 4.9 SEM micrographs
of **a** chitosan and **b** HA/
chitosan scaffolds after
in vitro degradation for
4 weeks (Sun et al. 2014)

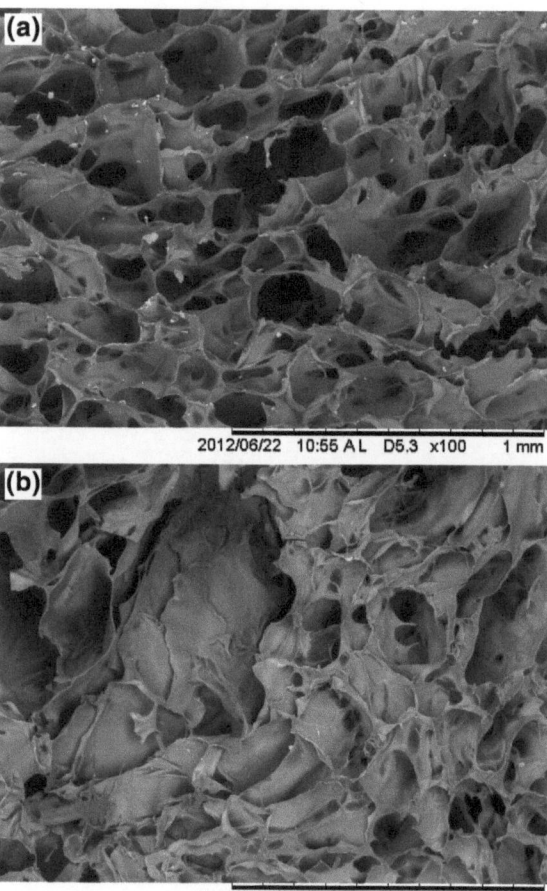

degradation tests. The pore distribution was found to be more irregular after
4 weeks in PBS. Some major morphological changes were detected for composite
scaffolds after immersion in PBS for 4 weeks including the elongated and large
pores, collapsed pore walls etc. It was not observed in the pure polymer scaffolds.

The chitosan and HA/chitosan composite scaffolds were successfully fabricated
using thermally induced phase separation technique. By varying the materials and
processing parameters, degree of porosity and pore sizes of the scaffolds could be
modulated. The scaffolds were highly porous, exhibited interconnected porous
structure with pore sizes from 20 to 350 μm. HA nanoparticles were successfully
incorporated which reduced the initial water uptake properties of the scaffolds and
also enhanced compressive mechanical properties. Biodegradable and biocompatible porous scaffolds were demonstrated to be suitable for in vitro cell culture and
are expected to allow cell attachment and tissue in-growth when used in vivo.

4.2 HA/PCL Composite Scaffolds Using Freeze-Drying Technique

Hydroxyapatite (HA) is biological apatite has close similarity of composition with human bone and teeth. Thus, it has been used in biomedical application in orthopedic and dentistry. The wet slurry of HA was successfully produced by mixing an acetone solution of calcium nitrate 4-hydrate with an aqueous solution of ammonium phosphate and ammonium carbonate with control pH of 11. The nano-emulsion was kept in freezer about one day and after that was kept in freeze drying machine about three days to obtain dry HA powder with low degree of agglomeration. The nanoparticles were studied under scanning electron microscopy (SEM). Energy Dispersive Xray (EDX) showed the spectrum of elements in HA with Ca/P ratio close to biological bone. The polycaprolactone (PCL) and hydroxyapatite/polycaprolactone (HA/PCL) composite scaffolds were produced using thermally induced phase separation (TIPS) technique. The scaffolds were studied under SEM and it was observed that both types of scaffolds had porous structures. The pore sizes of HA/PCL scaffold was slightly decreased compared to PCL scaffold. The HA nanoparticles were successfully produced and the PCL and HA/PCL scaffolds showed promises for bone tissue engineering application. Poly(caprolactone) (PCL) were supplied by *Sigma-Aldrich*. PCL had the molecular weight of 70,000–90,000. All the chemicals used in the research such acetic acid, 1,4 dioxane were analytical grade. Nanoemulsion process (Zhou et al. 2008) was employed in-house to produce nano-sized HA for composite scaffolds.

$Ca(NO_3)_2 \cdot 4H_2O$ (acetone solution) was mixed with $(NH_4)_2HPO_4$ and NH_4HCO_3 (aqueous solution) at pH 11 at molar ratio of $Ca^{2+}:PO_4^{3-}:CO_3^{2-}$. The mixed solution was stirred for about 0.5 min and then filtered using Millipore glass vacuum filtration, washed 3 times using ultra-pure deionized water. The precipitation was transferred into glass tube and was emulsion-freezing for overnight. The dry powder was obtained after freeze-drying using freeze-drying vessel (Labconco, USA) and was ready to analyze. A scanning electron microscope (SEM) was used to observe and study the morphology of nanoparticles of HA and scaffolds. Energy dispersive X-ray (EDX) was used to analyze the elements of HA.

The scaffolds were fabricated using TIPS technique. Briefly, 1 g of PCL was weighted using analytical balance and put into centrifuge tube containing 10 ml of 1,4-dioxane. The centrifuge is kept at 50 °C in water bath in order to obtain homogenous solution. Then, a minishaker was used to get the homogenous mixture. After the solution was mixed, it was transferred into glass tube and was put into the freezer for solidification for overnight. The tubes were then transferred into freeze-drying machine. The final scaffolds were stored in vacuum desiccator for characterization. The fabrication of HA/PCL composite scaffold had the similar procedure as that of PCL scaffold. The weighed HA powder was added into centrifuge tube containing PCL polymer solution.

The scaffold morphology was studied under SEM to observe the size and distribution of the pores within PCL and HA/PCL scaffold surface. The presence of the element of HA in PCL/HA composite scaffold was determined using EDX.

The nanoemulsion of HA was slightly milky white color as shown in Fig. 4.10a. Figure 4.10b shows that freeze-dried HA powders. HA powder had flow ability and had lower agglomeration. The properties of flowability were very important because the powder could be dispersed thoroughly in the matrix polymer causing high performance in biological and mechanical state (Zhou et al. 2008).

EDX analysis of HA particles is given in Fig. 4.11. The elemental analysis of HA obtained from EDX analysis is given in Table 4.2.

Figure 4.12 shows the PCL scaffold and HA/PCL composite scaffold produced by TIPS technique. Size and shape of the scaffolds were depended on the tubes that

Fig. 4.10 **a** Nanoemulsion (*left*) and pure water (*right*). **b** Freeze-dried powder of HA nanoparticles (Hassan et al. 2012)

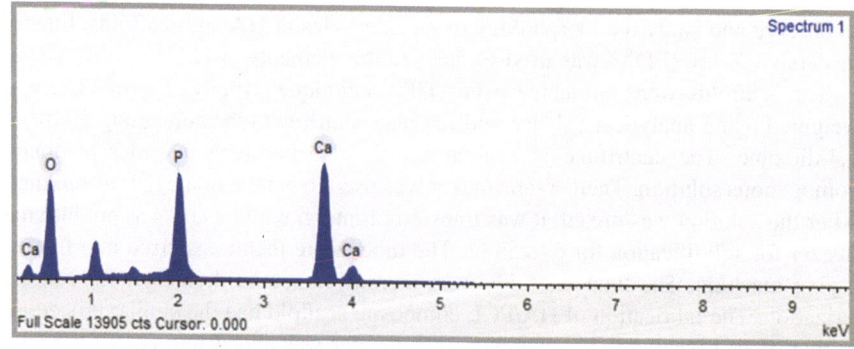

Fig. 4.11 An EDX spectrum of HA nanoparticles (Hassan et al. 2012)

Table 4.2 Quantification of element in HA nanoparticles (Hassan et al. 2012)

Element	Weight %	Weight % σ	Atomic %
Oxygen	61.483	0.239	78.345
Phosphorus	13.801	0.125	9.084
Calcium	24.715	0.178	12.571

Fig. 4.12 General appearance of PCL and PCL/HA scaffold (Hassan et al. 2012)

were used as shown in Fig. 4.13. Both PCL and HA/PCL scaffolds had sponge-like structures with good physical properties.

Figure 4.13a, b shows the morphology of PCL and HA/PCL scaffolds. The scaffolds had pore sizes from several microns to few hundred microns. The pores were totally interconnected. The pores showed open morphologies. In 10 % HA/PCL composite scaffold, HA nanoparticles were observed to be distributed within the pore walls of the scaffold. EDX analysis was performed to confirm the presence of HA in the pore walls of HA/PCL scaffold. It was observed that the pore sizes decreased in comparison to pure PCL scaffold with the incorporation of 10 % (w/w) HA nanoparticles. No significant agglomeration of HA nanoparticles was observed in the HA/PCL composite scaffold.

Using homogenizer at high speed reduced the agglomeration and also helped the HA nanoparticles to disperse in the polymer pore walls. The characteristic pore morphology was caused by solid-liquid phase separation and the heat transfer mechanism during freezing and freeze-drying process (Sultana and Khan 2012; Sultana and Wang 2008b, 2012; Sultana and Abdul Kadir 2011).

Fig. 4.13 SEM micrographs
of **a** PCL and **b** HA/PCL
scaffolds (Hassan et al. 2012)

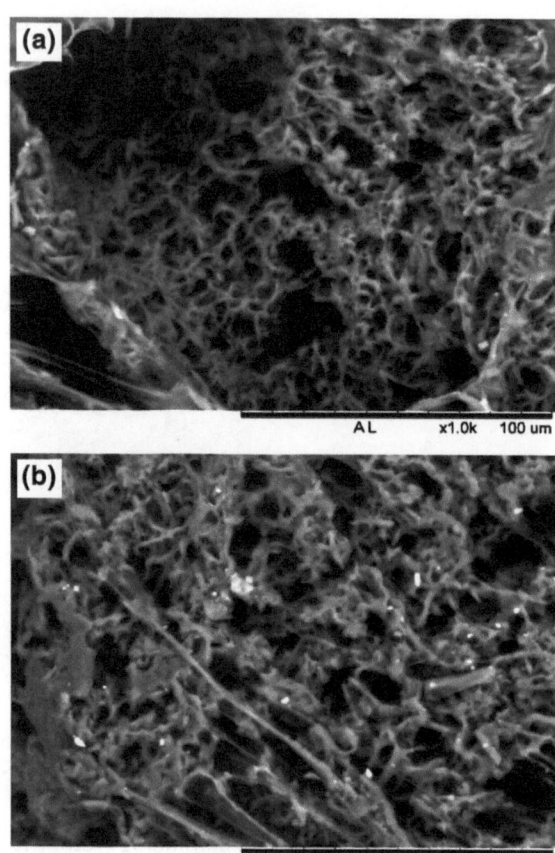

4.3 Conclusions

Nano-sized hydroxyapatite particles were produced using a nano-emulsion process.
Highly porous, three dimensional scaffolds with adequate distribution of pores were
fabricated using TIPS technique using PCL polymer and nano-sized HA. With the
incorporation of HA nanoparticle, composite HA/PCL scaffold exhibited smaller
pores compared to PCL scaffold. The PCL and HA/PCL scaffolds had suitable
properties for the bone tissue engineering application.

Acknowledgments The authors would like to thank Univesiti Teknologi Malaysia (UTM) for the
facilities and support. Authors also acknowledge FRGS Vote: 4F126, GUP (Tier 1) 03H13,
Ministry of Higher education (MOHE), RMC and UTM for financial support.

References

Atala, A., & Lanza, R. P. (2002). *Methods of tissue engineering*. San Diego, CA: Academic Press.

Chen, V. J., & Ma, P. X. (2005). *Scaffolding in tissue engineering*. Boca Raton: Taylor & Francis.

Gibson, L. J., & Ashby, M. F. (1997). *Cellular solids: Structure and properties*. Cambridge: Cambridge University Press.

Griffith, L. G. N. G. (2002). Tissue engineering-current challenges and expanding opportunities. *Science, 295*, 1009–1014.

Han, J., Zhou, Z., Yin, R., Yang, D., & Nie, J. (2010). Alginate chitosan/hydroxyapatite polyelectrolyte complex porous scaffolds: Preparation and characterization. *International Journal of Biological Macromolecules, 46*, 199–205.

Hassan, M. I., Mokhtar, M., Sultana, N., & Khan, T. H. (2012). Production of hydroxyapatite (HA) nanoparticle and HA/PCL tissue engineering scaffolds for bone tissue engineering (pp. 239–242). *IEEE*.

Hollister, S. J. (2005). Porous scaffold design for tissue engineering. *Nature Materials, 4*, 518–524.

Hsu, Y.-Y., Gresser, J. D., Trantolo, D. J., Lyons, C. M., Gangadharam, P. R., & Wise, D. L. (1997). Effect of polymer foam morphology and density on kinetics of in vitro controlled release of isoniazid from compressed foam matrices. *Journal of Biomedical Materials Research, 35*, 107–116.

Khan, Y., Yaszemski, M. J., Mikos, A. G., & Laurencin, C. T. (2008). Tissue engineering of bone: Material and matrix considerations. *Journal of Bone and Joint Surgery. American Volume, 90*, 36–42.

Lanza, R. P., Langer, R. S., & Vacanti, J. (2007). *Principles of tissue engineering*. Amsterdam, Boston: Elsevier/Academic Press.

Ma, P. X. (2004). Scaffolds for tissue fabrication. *Materials Today, 7*, 30–40.

Park, J. B., & Bronzino, J. D. (2003). *Biomaterials: Principles and applications*. Boca Raton: CRC Press.

Ratner, B. D. (2004). *Biomaterials science: An introduction to materials in medicine*. San Diego, CA, London, UK: Elsevier Academic Press.

Rinaudo, M. (2006). Chitin and chitosan: Properties and applications. *Progress in Polymer Science, 31*, 603–632.

Schugens, C., Maquet, V., Grandfils, C., Jerome, R., & Teyssie, P. (1996). Biodegradable and macroporous polylactide implants for cell transplantation: 1. Preparation of macroporous polylactide supports by solid-liquid phase separation. *Polymer, 37*, 1027–1038.

Sultana, N., & Abdul Kadir, R. (2011). Study of in vitro degradation of tissue engineering scaffolds based on biodegradable polymers. *African Journal of Biotechnology, 10*, 18709–18715.

Sultana, N., & Khan, T. H. (2012). In vitro degradation of PHBV scaffold and HA/PHBV composite scaffolds containing hydroxyapatite nanoparticle for bone tissue engineering. *Journal of Nanomaterials, 2012*, 1–12.

Sultana, N., & Wang, M. (2008a). Fabrication of HA/PHBV composite scaffolds through the emulsion freezing/freeze-drying process and characterisation of the scaffolds. *Journal of Materials Science Materials in Medicine, 19*, 2555–2561.

Sultana, N., & Wang, M. (2008b). PHBV/PLLA-based composite scaffolds containing nano-sized calcium phosphate particles for bone tissue engineering. *Journal of Experimental Nanoscience, 3*, 121–132.

Sultana, N., & Wang, M. (2012). PHBV/PLLA-based composite scaffolds fabricated using an emulsion freezing/freeze-drying technique for bone tissue engineering: Surface modification and in vitro biological evaluation. *Biofabrication, 4*(1), 015003.

Sun, T., Khan, T. H., & Sultana, N. (2014). Fabrication and in vitro evaluation of nanosized hydroxyapatite/chitosan-based tissue engineering scaffolds. *Journal of Nanomaterials, 2014*.

Tong, H.-W., Wang, M., Li, Z.-Y., & Lu, W. W. (2010). Electrospinning, characterization and in vitro biological evaluation of nanocomposite fibers containing carbonated hydroxyapatite nanoparticles. *Biomedical Materials, 5,* 054111.

Van Vlierberghe, S., Cnudde, V., Masschaele, B., Dubruel, P., De Paepe, I., & Jacobs, P. J. S. (2006). Porous gelatin cryogels as cell delivery tool in tissue engineering. *Journal of Controlled Release, 116,* e95–e98.

Vandevord, P. J., Matthew, H. W. T., Desilva, S. P., Mayton, L., Wu, B., & Wooley, P. H. (2002). Evaluation of the biocompatibility of a chitosan scaffold in mice. *Journal of Biomedical Materials Research, 59,* 585–590.

Wu, H.-D., Ji, D.-Y., Chang, W.-J., Yang, J.-C., & Lee, S.-Y. (2012). Chitosan-based polyelectrolyte complex scaffolds with antibacterial properties for treating dental bone defects. *Materials Science and Engineering C, 32,* 207–214.

Wu, H.-D., Yang, J.-C., Tsai, T., Ji, D.-Y., Chang, W.-J., Chen, C.-C., et al. (2011). Development of a chitosan-polyglutamate based injectable polyelectrolyte complex scaffold. *Carbohydrate Polymers, 85,* 318–324.

Zhou, W., Wang, M., Cheung, W., Guo, B., & Jia, D. (2008). Synthesis of carbonated hydroxyapatite nanospheres through nanoemulsion. *Journal of Materials Science: Materials in Medicine, 19,* 103–110.

Summary

This book aims to serve as a reference which combines the description and protocols for the promising field of tissue engineering and regenerative medicine. This book also focused on the fabrication of composite tissue engineering scaffolds through the electrospinning and freeze-drying processes. Evaluation of the scaffolds was also reported. The book focused on several areas: (1) description of scaffolding materials; (2) description of different processing route to fabricate scaffolds; (3) establishment of the scaffold fabrication technique using PCL polymers and HA by using electrospinning technique; (4) Fabrication of three dimensional composite scaffolds using PCL and HA via freeze-drying technique; (5) evaluation of the composite scaffolds. Successful scaffold fabrication can have vast potential in tissue engineering and regenerative medicine.

© The Author(s) 2015   61
N. Sultana et al., *Composite Synthetic Scaffolds for Tissue Engineering and Regenerative Medicine*, SpringerBriefs in Materials,
DOI 10.1007/978-3-319-09755-8